AF616002

Modern Pathology of AIDS and Other Retroviral Infections

Modern Pathology of AIDS and Other Retroviral Infections

Application of Contemporary Methods

Editors
Paul Racz, Hamburg;
Ashley T. Haase, Minneapolis, Minn.;
Jean-Claude Gluckman, Paris

106 figures, 5 color plates and 13 tables, 1990

Basel · München · Paris · London · New York · New Delhi · Bangkok · Singapore · Tokyo · Sydney

Drug Dosage

The authors and the publisher have exerted every effort to ensure that drug selection and dosage set forth in this text are in accord with current recommendations and practice at the time of publication. However, in view of ongoing research, changes in government regulations, and the constant flow of information relating to drug therapy and drug reactions, the reader is urged to check the package insert for each drug any change in indications and dosage and for added warnings and precautions. This is particularly important when the recommended agent is a new and/or infrequently employed drug.

Printed in Switzerland by Friedrich Reinhardt AG, Basel
ISBN 3–8055–5079–0

Contents

Preface

The recent spread of human immunodeficiency viruses throughout the world underscores the importance of retroviral infections and emphasizes the need for techniques which can deepen our understanding of their pathogenesis. Both new and older techniques such as electron microscopy, immunohistochemistry and the polymerase chain reaction have provided invaluable insights into the complex relationships between viruses and their host cells in slow infections, but one technique, in situ hybridization, has emerged as a particularly powerful technology for studies of retroviral pathogenesis because it combines detection of viral genomes and transcripts with morphological information. This volume on the molecular pathology of retroviral infections, which grew out of a workshop whose primary focus was *in situ hybridization in retroviral infections* (Hamburg, West Germany, November 17–18, 1988), treats this topic in the detail it deserves, but also contains chapters on the polymerase chain reaction, immunohistochemistry and electron microscopy to provide a broad and up-to-date overview of the field of molecular pathology of retroviral infections, written by individuals who have developed and applied these techniques in retroviral research.

The book opens with an overview of the pathogenesis of animal lentivirus infections viewed at the level of the individual cell harboring virus as revealed by in situ hybridization and immunocytochemical methods. Subsequent chapters are devoted to the use of different techniques, such as hapten-modified probe for in situ hybridization, polymerase chain reaction amplification, electron microscopy, immunocytochemistry and immunohistochemical analysis, to investigate various retroviral and non-retroviral models. The pathophysiological aspects of animal and human retroviral diseases are then discussed and compared, especially visna/maedi and HIV infections but also FeLV (which provided the model for searching for a retrovirus in AIDS) and SIV, currently the most important AIDS model system.

The editors wish to thank the authors who contributed to this volume, and the European Community Commission Program on AIDS through the 'Concerted action on the pathophysiology and immunology of HIV-related diseases' for sponsoring the Hamburg workshop and publication of its proceedings. We dedicate this volume to Dr. Kurt A. Körber on his 80th birthday, to recognize and gratefully acknowledge the role he has played in promoting research in Europe through the *Award for the Advancement of European Science* (General Secretary Dr. H. Gretz) which has generously supported many of the authors of this work.

Paul Racz
Ashley T. Haase
Jean-Claude Gluckman

Racz P, Haase AT, Gluckman JC (eds): Modern Pathology of AIDS and
Other Retroviral Infections. Basel, Karger, 1990, pp 1–16

Analysis of Lentivirus Infections by in situ Hybridization

Introduction to the Lentiviruses

Ashley T. Haase

Department of Microbiology, University of Minnesota, Minneapolis, Minn., USA

During the 1930s, sheep imported from Germany introduced diseases into Iceland such as maedi and visna which reached epidemic proportions in the ensuing decade. Maedi, an interstitial pneumonia, and visna, a subacute encephalomyelitis, were shown by Björn Sigurdsson to be transmitted by viruses that caused disease with incubation periods of months or years. This novel time frame was the basis for Sigurdsson's definition of a new class of slow infections, a designation which now extends to a taxonomically diverse group of infectious agents with the common properties of long incubation periods, protracted course, and generally fatal outcome. In the next two decades, the viruses responsible for visna and maedi were shown to be members of the family of retroviridae, in a separate subfamily, the lentivirinae, or slow retroviruses, named for the distinguishing feature of the infections they cause [1]. With the discovery that the cause of the acquired immunodeficiency syndrome (AIDS), human immunodeficiency virus (HIV), is a lentivirus [2, 3], there has been renewed interest in the lentivirinae and an expansion in the membership to include immunodeficiency viruses of other primates, ruminants and cats (fig. 1). This introductory chapter is devoted to a discussion of lentivirinae viewed at the level of individual cell-virus interactions through the use of in situ hybridization methodologies. This special perspective on animal lentivirus infections has already provided invaluable insights into mechanisms of lentivirus persistence and dissemination in the face of host defenses, and the slow tempo of the disease process.

The Basis for the Persistence and Dissemination of Lentiviruses: Restricted Viral Gene Expression

All lentiviruses, including HIV, persist and are disseminated in and between infected individuals despite virus-specific neutralizing antibody and cytotoxic T cells [4–6]. I suggested a number of years ago that the explanation for the inability of the host to clear infection is the restricted nature of viral gene expression in vivo where diminished synthesis and presentation of viral antigens accounts for ineffective detection and destruction by immune surveillance mechanisms [7]. This results in a population of infected cells which act as a reservoir to perpetuate infection. The covert infection of a mobile cell population as, for example, lymphocytes or monocytes, was advanced as the explanation for continued dissemination of the infectious agent despite host defenses [8].

The Rationale for in situ Hybridization: Restricted Gene Expression in the Focal Nature of Infection

These hypotheses predict stable association of the lentiviral genome with its host under conditions where constraints on gene expression make the cell a poor target for the immune system. Testing these predictions turned out to be quite challenging and led us to introduce substantial improvements in the methods for the analysis of viral infections in single cells, particularly in situ hybridization [9, 10]. The reason for this is the focal nature of infection and restriction in gene expression. If only 0.1–1% of the population harbors one or a few viral genomes in a repressed state, extraction of macromolecules from the population will result in dilution from uninfected cells of viral DNA, RNA or protein, to the point where they will go undetected. In situ hybridization, immunofluorescence and cytochemistry, on the other hand, can reveal viral genes and their products in single infected cells in a tissue section with thousands of uninfected cells.

The Visna Paradigm

The realization that single cell technologies were needed if we were to understand the pathogenesis of slow infection emerged from studies of visna virus in sheep. When we extracted DNA from tissues of infected animals from

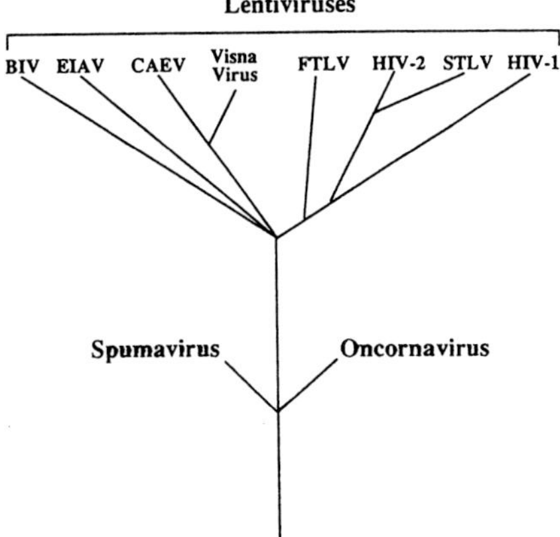

Fig. 1. Taxonomic relationships of retrovirus subfamilies. Adapted from Ho et al. [29]. BIV = Bovine immunodeficiency virus; EIAV = equine infectious anemia virus; CAEV = caprine-arthritis encephalitis virus; FTLV = feline T lymphotropic virus; STLV = Simian T lymphotropic virus; HIV-1, 2 = human immunodeficiency virus types 1 and 2.

which virus could be recovered by implanting tissue fragments, we failed to detect viral DNA using sensitive methods that had already proved successful in productive infections in tissue culture in demonstrating a DNA intermediate in the visna virus life cycle [11]. This led to the correct conclusion that the failure to detect viral sequences was a consequence of dilution of viral sequences in the isolation of DNA, and provided a strong motive to systematically investigate the parameters that determine the sensitivity and specificity of in situ hybridization. By adjusting the conditions of fixation and subsequent treatment of the tissues, to enhance diffusion of virus-specific probes to their target sequences in the cell, we initially achieved sensitivities in the range of 10–100 copies of viral DNA per cell with a 4-month exposure [7]. This proved sufficient to identify foci of cells in the brain of an infected sheep with visna DNA at a frequency which far exceeded the antigen-positive cells detected by immunofluorescence in subjacent sections. These findings fulfilled the predictions of the restricted expression hypothesis and illustrate the power of the in situ hybridization technologies to address major issues of pathogenesis in vivo.

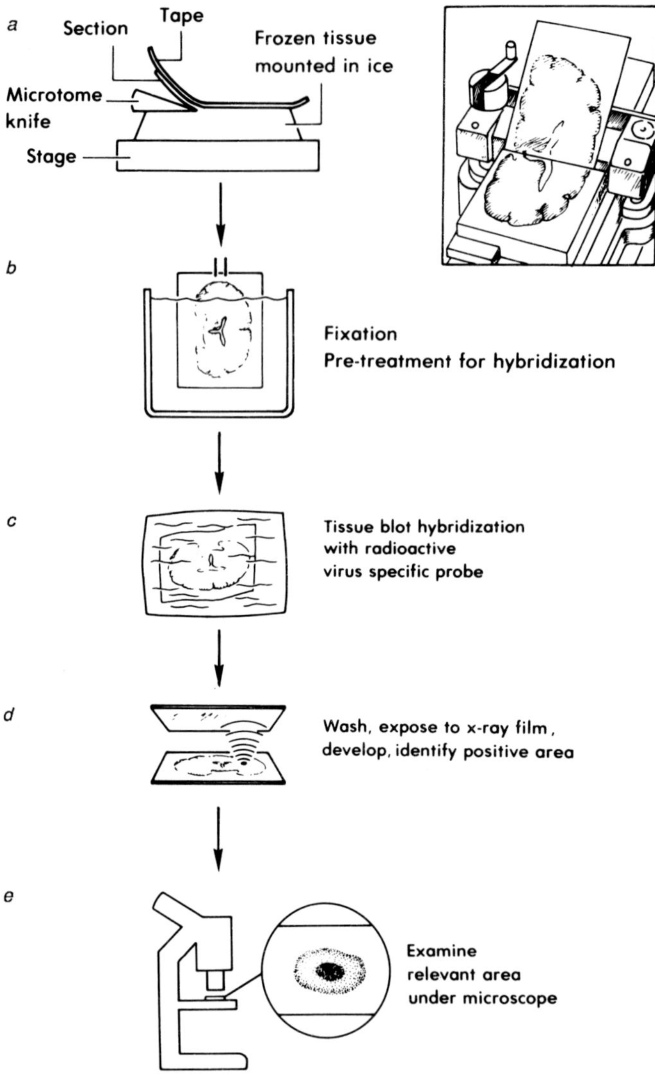

Fig. 2. Principles and major steps in combined macroscopic-microscopic hybridization. Explanation given in the text. Reprinted with permission from Haase et al. [13].

Further Refinements in Sensitivity, Sampling Power, and Quantitation

In the period following demonstration of the visna provirus in vivo, Michel Brahic and I continued to work on improving in situ hybridization detection of viral DNA and RNA, and developed protocols (Appendix, p. 13) which brought the sensitivity into the range of a few copies of visna DNA or RNA per cell [9, 10]. For RNA hybridizations, in addition to the HCl, protease, and high temperature treatmtents to enhance diffusion, we increased access of probe to target by reducing its size and using it in vast excess, both to increase diffusion and to drive the reaction to completion in a reasonable time (12–60 h, depending on the complexity and number of copies of viral sequence). For RNA hybridizations we also used complementary single-stranded DNA probes (now largely superseded by single-stranded RNA probes) [12] to circumvent the competing reassociation reaction of double-stranded probes. Double-stranded probes were used for DNA hybridizations, in combination with macromolecular crowding reagents such as dextran sulfate or polyethylene glycol, to build up at the target sequence a network of complementary radiolabelled strands and thus increase the signal at this site.

Optimization of the hybridization techniques and the introduction of ^{35}S and ^{125}I precursors, which increase both the specific activity of the probes (to about 10^9 dpm/μg or more) and the efficiency of latent image formation (from about 0.1 grain per ^3H disintegration to 0.2 and 0.5 respectively for ^{125}I and ^{35}S), have extended sensitivities into the attogram range, or the equivalent of a few copies of a 2 kb mRNA per cell [9, 10]. ^{125}I-labelled probes also can be used to enhance the sampling power of in situ analysis of lentivirus gene expression as shown in figure 2 and 3. In this combined macroscopic-microscopic screening technique [13], large frozen sections of brain, other organs, or small animals are cut and transferred to clear tape or large glass slides. After pretreatments, the sections are hybridized to ^{125}I-labeled probes, washed, and placed against X-ray film. The gamma emission of ^{125}I will record at low resolution the location of viral genes in the large section. The section is then coated with nuclear track emulsion, developed after a few days, and areas with viral genes, identified by aligning the section with the X-ray film, are then examined microscopically. The grains over cells with viral genes are the result of the Auger electrons emitted by ^{125}I which are of low energy like ^3H and thus give a well-localized signal. An example of the use of this technique to document the paraventricular spread of visna virus is shown in figure 3. With this technique, tissues equivalent to several hundred small sections can be surveyed for viral genes without sacrificing cellular resolution.

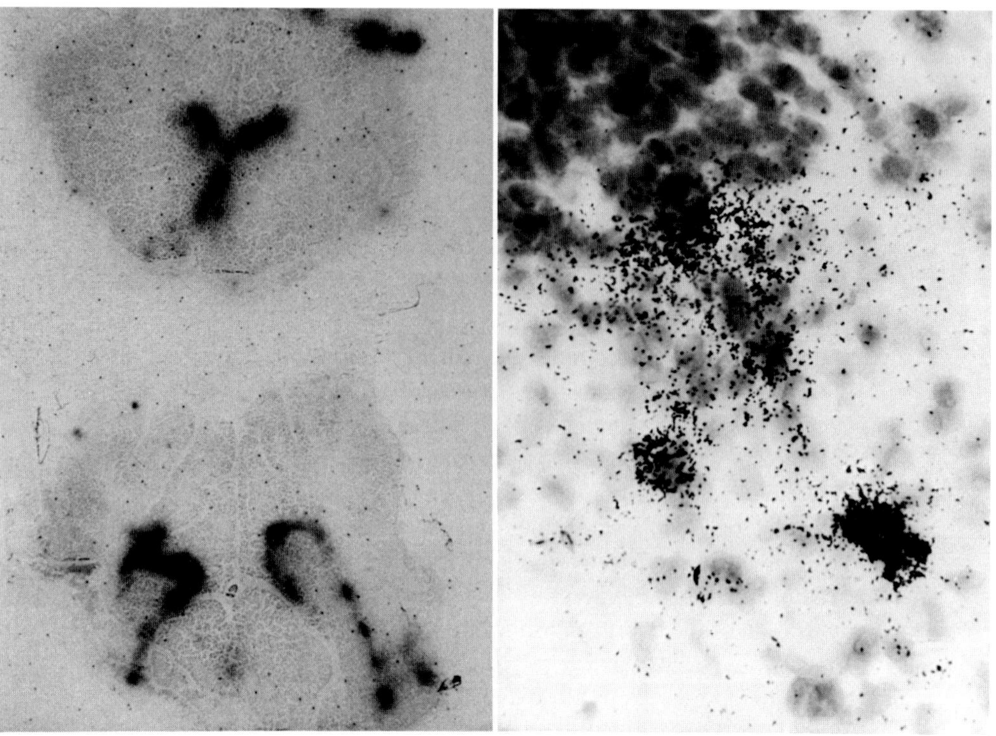

Fig. 3. Example of the use of macroscopic-microscopic hybridization to demonstrate the periventricular spread of visna virus. *a* The darkened areas represent the signal generated from virus-specific probe hybridized to viral RNA and diluted by pressing the larger brain sections against X-ray film. *b* The cells generating the signal are detected by radioautography with nuclear track emulsion. Reprinted with permission from Haase et al. [13].

Radioisotopic probes and radioautography also yield quantifiable signals so that the number of copies of viral DNA or RNA per cell can be estimated. In practice, this is accomplished by infecting ovine cells that will support productive infection and then collecting cells in the course of infection where the number of copies will vary over three orders of magnitude (see Appendix). The average copy number is determined in one aliquot at each time point by solution or blot hybridization; the average grain count over background is measured in another aliquot by in situ hybridization. In this way, a relationship is established between the number of grains and the number of copies per cell, valid over a wide range of copy numbers.

The Lysogeny Hypothesis

While the mechanism of persistence in visna had been established, by showing that the genome was present in many cells that had no detectable viral antigens, these experiments did not identify the stage in the viral life cycle where the restriction in gene expression is effected. This issue was addressed with the sensitive and quantitative methods of in situ hybridization just described. By comparing the number of copies of visna RNA per cell in vivo, in cells in choroid plexus [14], in glial cells [15], and in alveolar macrophages [16], with productively infected tissue culture cells, we found in each case that the accumulation of viral RNA was reduced by about two orders of magnitude in vivo. Because the block in gene expression occurs primarily at the level of transcription (or RNA stability) we initially referred to the mechanism of persistence as lysogeny on a grand scale, by analogy to the lysogenic relationship of the temperate bacteriophages to their bacterial hosts [17]. Because visna virus may well replicate by using extrachromosomal DNA as the template for transcription, it is more likely now that the mechanism is closer to pseudolysogeny [18].

The Trojan Horse Theory of Spread

Visna virus can also be recovered from blood and other tissue fluids throughout the course of infection despite the neutralizing antibody and virus-specific lymphocytes in these same fluids [4, 5]. While antigenic variation has been invoked to account the inability of the host to eradicate virus [19], this seems unlikely to hold, at least for visna virus [20, 21]. In my view, restricted gene expression accounts for both persistence and spread of lentiviruses [1]. For dissemination, it is the covert relationship of visna virus to the monocyte which provides the mechanism for the virus to circulate inside cells and go undetected by host defenses [8]. By analogy to the breaching of the defenses of Troy, I have referred to this mode of spread as the Trojan horse mechanism [8].

Experimental support for the Trojan horse theory again comes largely from analyses of tissue fluids by quantitative in situ hybridization. We studied cells in cerebrospinal fluid responding to previously inoculated virus, using a self-sealing reservoir from which we could repeatedly sample cells and fluid. Under these conditions, infected cells are enriched 10- to 100-fold compared to blood. At a time when the CSF had neutralizing antibody, we recovered

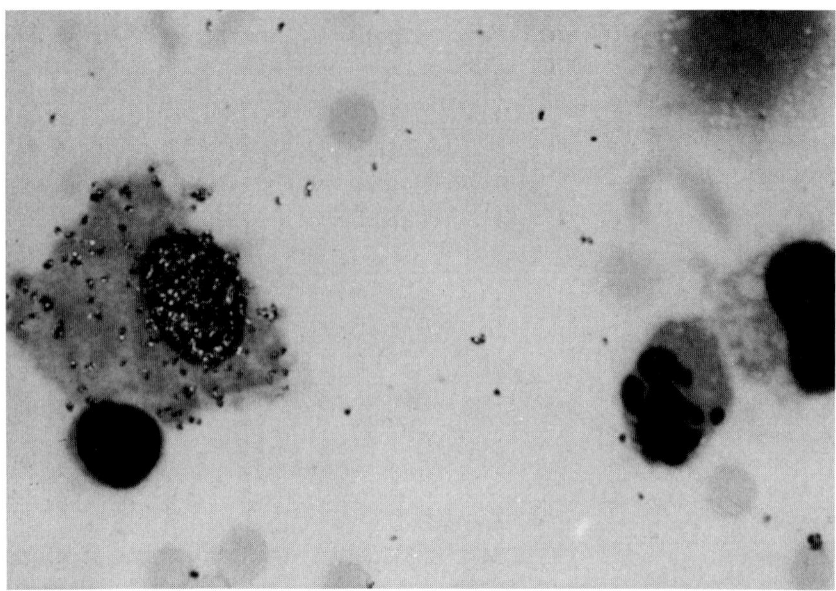

Fig. 4. The monocyte as Trojan horse. The figure shows visna RNA primarily in the nucleus of a monocyte detected by in situ hybridization and radioautography. The adjoining lymphocyte and neurotrophic granulocyte in the field, identified by their distinctive morphology, do not contain viral RNA. The infected cell harbors the visna genome in cerebrospinal fluid containing neutralizing antibody. Reprinted with permission from Peluso et al. [8].

virus-infected monocytes with visna RNA at restricted levels (fig. 4). These cells, however, could transfer infectivity to a monolayer of permissive cells cultured in vitro. These findings are in full accord with the predictions of the Trojan horse model of lentivirus dissemination.

Spectrum of States of Lentivirus Gene Expression and the Tissue Lesions

The lysogeny model is also imprecise in another sense: the lentivirus genome, we have discovered, is never really transcriptionally silent. The average copy number may be quite low (50–100 copies of RNA per cell) but some RNA is always detectable, and there is a range which extends to nearly permissive levels of a thousand copies of RNA per cell or more [15, 16]. This spectrum or gradient of gene expression was discovered with the quantitative methods of in situ hybridization described in the preceding sections [9, 10]. The high copy number end of the spectrum also correlated precisely with the extent of inflammation, accompanying tissue lesions, and the detection of viral antigens in subjacent sections. Collectively, these observations argue for an immunopathological process [22] provoked and sustained by the infected cells in the spectrum of gene expression which express inciting viral antigens. Thus, in visna as in other slow infections [23, 24], persistence can be attributed to the infected cell with the greatest repression in gene expression, and disease to a process set in motion by the host's response to cells with higher levels of gene expression. The slow tempo is a consequence of the small fraction of cells in which these events take place over time.

Combining in situ Hybridization and Immunocytochemistry to Understand the Specificity of the Pathological Process

There is some specificity in the immunopathological process in visna, which is predominantly directed against white matter. Indeed, in some cases primary demyelination has been documented which closely resembles the human demyelinating disease, multiple sclerosis [25]. We have devised and used a combined immunocytochemical in situ hybridization technique [26] to show that the explanation, at least in part, for the white matter lesions is that the oligodendrocyte, the cell responsible for myelination in the CNS, is infected, and thus is plausibly a target for the immunopathological response.

In the combined assay, antibodies specific for viral or cellular proteins (in this case antibodies to sheep oligodendrocytes) are reacted with the tissue sections and the antibodies complexed with their antigens are subsequently detected by peroxidase-labelled reagents and substrates such as diaminobenzidine (DAB) or phenylenediamine-pyrocatechol (PDD). Polymers formed in this way mark the oligodendrocyte (brown or black stain) and are sufficiently stable to withstand the subsequent treatments for in situ hybridization with virus-specific probes. In the developed radioautographs, an increased number of grains over background in a stained cell unequivocally establishes infection of the oligodendrocyte. With this technique we documented infection of the oligodendrocyte in an animal with advanced symptoms of the encephalomyelitis induced by visna virus [15].

The oligodendrocyte is only one of the cell types in which visna virus infection has been demonstrated with the combined assay. We have also used antibodies to glial fibrillary acidic protein (GFAP), an astrocyte-specific anti-

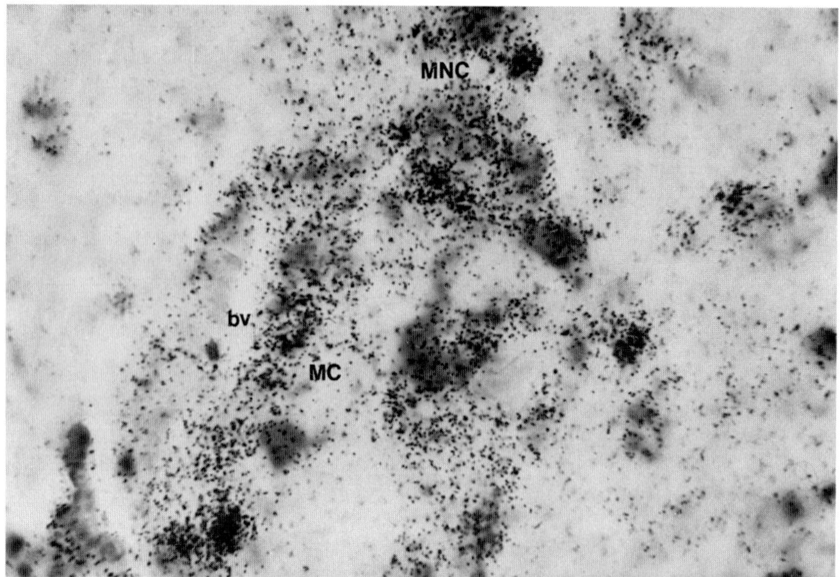

Fig. 5. HIV RNA demonstrated in the nervous system of an individual with the AIDS dementia complex. See text for explanation. MNC = Multinucleated cell; bv = blood vessel; MC = monocyte.

gen, to show that the host range of visna virus includes this glial element as well. This result distinguishes visna virus infection of the CNS from HIV where we have been unable to detect either HIV DNA or RNA in tissue sections from individuals who had the AIDS dementia complex [M. Zupancic et al., unpubl. data]. We find, as others have reported [27, 28], HIV RNA primarily in the monocytes, macrophages, and nucleated giant cells (fig. 5) in the perivascular inflammatory infiltrates, or in the resident macrophages of the CNS, the microglia.

Conclusions and Prospects

In this chapter, I have briefly chronicled some of the highlights of the investigations of pathogenesis of animal lentivirus infections in which the techniques of in situ hybridization and immunocytochemistry have deepened our understanding of these slow infections by bringing to light the often subtle and complex relationship between viruses and their host cells. The ensuing chapters expand on this theme, with illustrations of the power of these techniques in HIV and other viral infections to address important issues in pathogenesis. These later chapters also record the contemporary advances in these methods that raise the sensitivity and precision of diagnosis of lentivirus infection to unprecedented levels. It is on these foundations, and the great efforts underway to develop vaccines and effective antiviral agents, that we can hope to contain the epidemic caused by the first human lentivirus.

References

1 Haase, A. T.: The pathogenesis of lentivirus infections. Nature *322:* 130–136 (1986).

2 Sonigo, P.; Alizon, M.; Staskus, K.; Klatzmann, D.; Cole, S.; Danos, O.; Retzel, E.; Tiollais, P.; Haase, A.; Wain-Hobson, S.: Nucleotide sequence of the visna lentivirus: relationship to the AIDS virus. Cell *42:* 369–382 (1985).

3 Gonda, M. A.; Wong-Staal, F.; Gallo, R. C.; Clements, J. E.; Narayan, O.; Gilden, R. V.: Sequence homology and morphologic similarity of HTLV-III and visna virus, a pathogenic lentivirus. Science *227:* 173–177 (1985).

4 Pétursson, G.; Nathanson, N.; Georgsson, G.; Pantich, H.; Pálsson, P. A.: Pathogenesis of visna. I. Sequential virologic, serologic, and pathologic studies. Lab. Invest. *35:* 402–412 (1976).

5 Griffin, D. E.; Narayan, O.; Adams, R. J.: Early immune responses in visna, a slow viral disease of sheep. J. Infect. Dis. *138:* 340–349 (1978).

6 Fauci, A. S.: The human immunodeficiency virus: Infectivity and mechanisms of pathogenesis. Science *239:* 617–622 (1988).

7 Haase, A. T.; Stowring, L.; Narayan, O.; Griffin, D.; Price, D.: Slow persistent infection caused by visna virus: role of host restriction. Science *195:* 175–177 (1977).

8 Peluso, R.; Haase, A.; Stowring, L.; Edwards, M.; Ventura, P.: A Trojan horse mechanism for the spread of visna virus in monocytes. Virology *147:* 231–236 (1985).

9 Haase, A. T.; Brahic, M.; Stowring, L.: Detection of viral nucleic acids by in situ hybridization; in Maramorosch, K., Koprowski, H. (eds): Methods in Virology, vol. VII, pp. 189–226 (Academic Press, New York 1984).

10 Haase, A. T.: Analysis of viral infections by in situ hybridization; in Valentino, K., Roberts, J., Barchas, J. (eds.): In situ Hybridization – Applications to Neurobiology. Symposium Monograph, pp. 197–219 (Oxford University Press, Fairlawn, N.J. 1986).

11 Haase, A. T.; Varmus, H. E.: Demonstration of a DNA provirus in the replication of the slow virus visna. Nat. New Biol. *245:* 237–239 (1973).

12 Cox, K. H.; DeLeon, D. V.; Angerer, L. M.; Angerer, R. C.: Detection of mRNAs in sea urchin embryos by in situ hybridization using asymmetric RNA probes. Dev. Biol. *101:* 485–502 (1984).

13 Haase, A. T.; Gantz, D.; Blum, H.; Stowring, L.; Ventura, P.; Geballe, A.; Moyer, B.; Brahic, M.: Combined macroscopic and microscopic detection of viral genes in tissues. Virology *139:* 201–206 (1985).

14 Brahic, M.; Stowring, L.; Ventura, P.; Haase, A. T.: Gene expression in visna virus infection. Nature *292:* 240 (1981).
15 Stowring, L.; Haase, A. T.; Pétursson, G.; Georgsson, P.; Pálsson, P.; Lutley, R.; Roos, R.; Szuchet, S.: Detection of visna virus antigens and RNA in glial cells in foci of demyelination. Virology *141:* 311–318 (1985).
16 Geballe, A. P.; Ventura, P.; Stowring, L.; Haase, A. T.: Quantitative analysis of visna virus replication in vivo. Virology *141:* 148–154 (1985).
17 Haase, A. T.: The slow infection caused by visna virus. Curr. Top. Microbiol. Immunol. *72:* 101–156 (1975).
18 Harris, J. D.; Blum, H.; Scott, J.; Traynor, B.; Ventura, P.; Haase, A.: Slow virus visna: Reproduction in vitro of virus from extrachromosomal DNA. Proc. Natl. Acad. Sci. USA *81:* 7212–7215 (1984).
19 Narayan, O.; Griffin, D. E.; Chase, J.: Antigenic shift of visna virus in persistently infected sheep. Science *197:* 376–378 (1977).
20 Thormar, H.; Barshatzky, M. R.; Arnesen, K.; Kozlowski, P. B.: The emergence of antigenic variants is a rare event in long-term visna virus infection in vivo. J. Gen. Virol. *64:* 1427–1432 (1983).
21 Lutley, R.; Pétursson, G.; Pálsson, P. A.; Georgsson, G.; Klein, J.; Nathanson, N.: Antigenic drift in visna: virus variation during long-term infection of Icelandic sheep. J. Gen. Virol. *64:* 1433–1440 (1983).
22 Nathanson, N.; Panitch, H.; Pálsson, P. A.; Pétursson, G.: Pathogenesis of visna. II. Effect of immunosuppression upon early central nervous system lesions. Lab. Invest. *35:* 444–451 (1976).
23 Haase, A. T.: The pathogenesis of slow virus infections: Molecular analyses. J. Infect. Dis. *153:* 441–447 (1986).
24 Haase, A. T.: In situ hybridization and covert virus infections; in Notkins, A. L., Oldstone, M. B. A. (eds.): Concepts in Viral Pathogenesis II, pp. 310–316 (Springer, Berlin 1986).
25 Georgsson, G.; Martin, J. R.; Klein, J.; Pálsson, P. A.; Nathanson, N.; Pétursson, G.: Primary demyelination in visna: An ultrastructural study of Icelandic sheep with clinical signs following experimental infection. Acta Neuropathol. (Berl.) *57:* 171–178 (1982).
26 Brahic, M.; Haase, A. T.; Cash E.: Simultaneous in situ detection of viral RNA and antigens. Proc. Natl. Acad. Sci. USA *81:* 5445–5448 (1984).
27 Gartner, S.; Markovits, P.; Markovitz, D. M.; Kaplan, M. H.; Gallo, R. C.; Popovic, M.: The role of mononuclear phagocytes in HTLV-III/LAV infection. Science *233:* 215–219 (1986).
28 Koenig, S.; Gendelman,H. E.; Orenstein, J. M.; Dal Canto, M. C.; Pezeshkpour, G. H.; Yungbluth, M.; Janotta, F.; Aksamit, A.; Martin, M. A.; Fauci, A. S.: Detection of AIDS virus in macrophages in brain tissue from AIDS patients with encephalopathy Science *233:* 1089–1093 (1986).
29 Ho, D. D.; Pomerantz, R. J.; Kaplan, J. C.: Pathogenesis of infection with human immunodeficiency virus. N. Engl. J. Med. *317:* 278–286 (1987).

Ashley T. Haase, MD, Department of Microbiology, School of Medicine,
University of Minnesota, 1460 Mayo Memorial Building, Box 196 UMHC,
420 Delaware Street S.E., Minneapolis, MN 55455-0312 (USA)

Appendix: Protocols[1]

(1) Flow diagram of procedures for quantitative in situ hybridization.

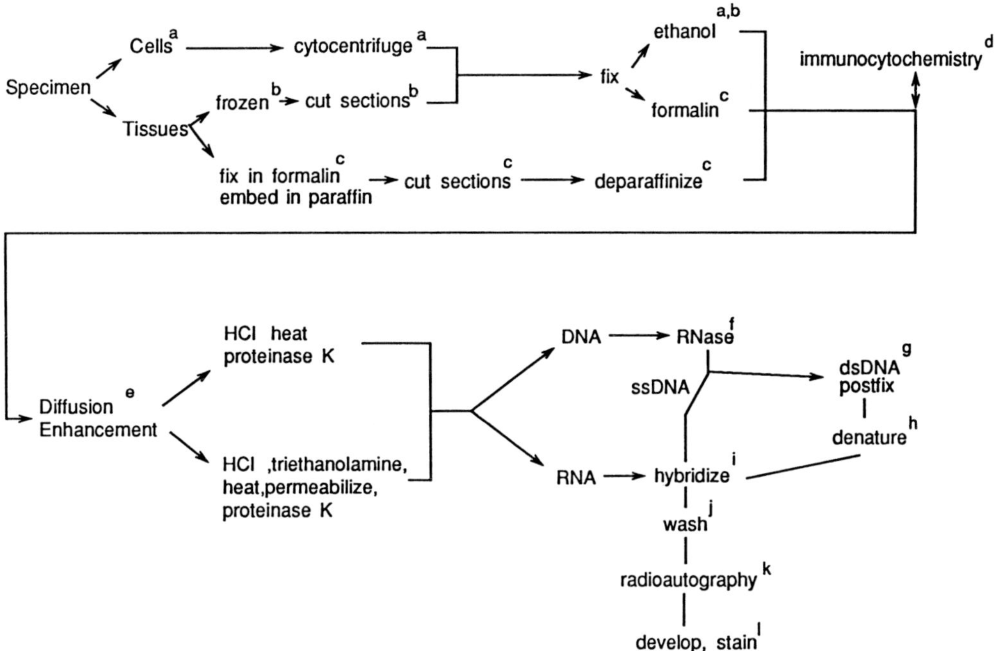

a Remove cells in monolayers with trypsin EDTA at 37°C for 5 min. Centrifuge at 400 *g* for 5 min. Resupend in PBS lacking Ca^{2+} and Mg^{2+}. Recentrifuge and resuspend in PBS at 10^6 cells/ml. Cytocentrifuge at 400 rpm for 5 min. Use washed and acetylated glass slides coated with DM. Air dry. Fix for 15 min in 3:1 ethanol/acetic acid, and for 5 min in ethanol at room temperature. Store desiccated at 4°C.

b Freeze 0.5- to 1.0-cm tissue sections between two blocks of dry ice. Store at –70°C. Cut into 4- to 10-μm sections. Air dry and fix in ethanol solutions or formalin for 20 min, as in c. Store in 70% ethanol.

c Perfuse animals with formalin solutions, or put tissue samples in 10% buffered formalin or (freshly prepared) 4% paraformadehyde for 2 h. Transfer and store at 4°C in 70% ethanol. Paraffin embed, cut into 5-μm section, and pick up the sections on treated glass slides coated with Elmer's glue. Alternatively, sprinkle gelatin on top of water at 42°C

[1]Reprinted with permission from Haase [10].

and pick the sections up; the gelatin will improve the adherence of the sections. Heat the slides at 37 °C overnight and for 3 h at 60 °C. Deparaffinize in xylene (three 5-min washes in xylene); wash twice in ethanol. Rehydrate through 90 and 70% ethanol and PBS before pretreating.

d Immunocytochemistry. Fix frozen sections in cold ethanol (4 °C) for 20 min, or use aldehyde-fixed material. Rehydrate and permeabilize in cold PBS with 0.01% Triton X-100 for 4 min. Wash twice in PBS at 4 °C, for 5 min each. React sequentially with normal serum diluted 1:200 in PBS (20 min at room temperature); monospecific antisera diluted usually 1:100 to 1:300 in normal serum-PBS (30 min at room temperature); 1:200 biotinylated antispecies IgG (30 min at room temperature); 0.3% H_2O_2 in methanol for 30 min (to block endogenous peroxidase in monocytes, etc.), avidin biotinylated horseradish peroxidase complex (ABC reagent, Vector Labs) for 30 min; 0.2% H_2O_2 with 0.5 mg/ml phenylene-diamine HCl, 1 mg/ml pyrocatechol, or 0.5 mg/ml DAB in 0.1 *M* Tris-HCl, pH 7.6, for 5–10 min at room temperature. Wash the slides in PBS for 5 min between steps and in water after reacting in DAB or PPD. Refix in ethanol, pretreat for in situ hybridization, and acetylate (0.25% acetic anhydride in triethanolamine buffer, pH 8.0) prior to hybridization. After development, stain with 0.5% $CuSO_4$, 1% methyl green as for color microradioautography.

e Place slides in 0.2 *N* HCl for 20 min, dip in water, or in 150 mmol/l of triethanolamine for formalin-fixed material. Incubate the slides at 70 °C in 2 × SSC (SSC = 0.15 mol/l of NaCl; 0.015 mil/l of sodium citrate), dip in water, and permeabilize in 0.005% digitonin in 125 mmol/l of sucrose, 60 mmol/l of KCl, and 3 m*M*/l of HEPES, pH 7.2. Dip the slides in water and transfer to a solution of 20 mmol/l of Tris-HCl, pH 7.4, 2 mmol/l of $CaCl_2$, 5 μg/ml of a proteinase K. Digest at 37 °C for 15 min. Wash the slides twice in water, for 5 min each time, and dehydrate (for 5 min each, two washes in 70% ethanol and once in 95% ethanol), and air dry.

f For DNA measurements in cells that also contain viral RNA, the RNA must be removed by enzymatic treatment. Place the slides in 100 μg/ml of DNAse-free ribonuclease A and 10 μg/ml of ribonuclease T_1 in 2 × SSC, at 37 °C for 30 min. Digest in 20 μl under a coverslip in a humidified chamber. Remove the coverslip by washing the slide for 5 min in a freshly prepared solution of 0.1% diethylpyrocarbonate in 2 × SSC. Wash in 2 × SSC for 5 min and dehydrate (ssDNA), or postfix double-stranded DNA.

g Postfix dsDNA in freshly prepared paraformaldehyde. (Add 10 *N* NaOH dropwise to 5 g of paraformaldehyde, stirring in 100 ml of PBS. When the solution clears, neutralize with HCl and filter.) Place the slides in paraformaldehyde for 2 h in the dark, wash twice in 2 × SSC, dip in water, and denature.

h Place slides in 95% formamide 0.1 × SSC at 65 °C for 15 min. Quickly transfer to 0.1 × SSC at 4 °C for 2 min, dip in water, and dehydrate.

i Concentrate probe by ethanol precipitation with 10 μg of DNA carrier used in the hybridization solution (e.g., calf thymus DNA reduced in size to about 300 by sonication and depurination followed by base treatment). Estimate the number of genes; use 0.3 ng of probe for 1–10 copies of a 3- to 10-kb gene per cell; 0.7 ng for 10–100 copies; 15 ng for 1,000 copies or more. The total radioactivity should not exceed 2×10^5 dpm/slide. Resuspend the probe (5 μl/slide) in hybridization solution containing 50% deionized formamide, 0.6 mol/l of NaCl, 20 mmol/l of HEPES, pH 7.2, 1 m*M* EDTA, 1 × DM, 100 μg/ml of calf thymus DNA, 50 μg/ml of poly-A, 5% PEG (~8,000 molecular weight), and 100 μ*M* ATA. Denature the probe by boiling it for 30 s and quench in ice. For ^{35}S probes add DTT to 10 mm. Add solution to cells or section, cover with a clean siliconized glass coverslip, and seal with rubber cement. Hybridize for 60–70 h at 25–37 °C in the dark. [j]Remove the coverslip by breaking the seal with the tip of a scalpel or needle. Wash the slides twice for 5 min in hybridization wash solution (HWS; 50% formamide, 0.6 mol/l of NaCl, 1 mmol/l of EDTA, 10 m*M* phosphate buffer, pH 6). Wash in 2 × SSC for 1 h at 55 °C. Wash twice at room temperature for 2 h in 1 liter of HWS. Dehydrate in graded ethanols with 0.3 *M* ammonium acetate (to stabilize hybrids). For long exposure, wash for 3 days in 1.5 liter of HWS with scraps of nitrocellulose paper or Zetabind (Biorad).

k In total darkness, dip the dried slide in Kodak NTB-2 (or equivalent) nuclear track emulsion diluted 1:1 with 0.6 ammonium acetate. Dry for 2 h, and place the slides in a light-tight box with a small amount of a drying agent. Wrap the boxes with aluminium foil and store at 4 °C.

l Warm the boxes to room temperature for 1 h. Develop in total darkness. Use Kodak D19 developer for 3 min at room temperature, dip in water, in fixer for 3 min, wash twice in water, 1 min each time, then dry. Stain for 25 min in filtered Meyer's hematoxylin. Wash in distilled water for 15 min. Stain for 6 min in 0.25% eosin and 80% ethanol. Dehydrate through graded alcohols. Air dry. Leukocytes should be stained with May-Grünwald/Giemsa: Stain the RNA slides for 30 min in May-Grünwald. Stain the DNA slides for 45 min in May-Grünwald. (May-Grünwald is made up as a 0.3% solution in methanol, heated to 56 °C for 1 h, cooled to room temperature, and stirred overnight. Filter and dilute 1:1 with 0.01 *M* phosphate buffer, pH 6.8, to make a working solution.) Dip in phosphate buffer for 1 min, dip in water, and stain for 45 min to 1 h in Giemsa. (To make a working solution, dilute in phosphate buffer MCB concentrated stock 4–100 ml.) Decolorize in distilled water until the slides begin to turn from purple to blue. For color development of nuclear track emulsions, convert the black sliver grains after development in D19 to magenta or cyan-colored grains as follows: harden the emulsion by immersing the slide for 3 min in 0.37% formalin buffered with 0.5% Na_2CO_3. Wash for 2 min in tap water at room temperature, and bleach for 1 min in 10% $K_3Fe(CN)_6$, 5% KBr. Wash again in tap water, and develop for 1 min in freshly prepared dye coupler. (For magenta grains, add 100 mg of Eastman Kodak M-38 to 5 ml of ethanol with 0.2 *N* NaOH. Mix with 45 ml of a solution containing Eastman Kodak developer CD2, 100 mg of Na_2SO_3, 50 mg of KBr, and 1 g of Na_2CO_3. For cyan grains, use Eastman Kodak cyan dye coupler C-16.) Wash, bleach, and wash again, and then fix for 5 min in Na_2-S_2O_3 (24 g/100 ml of water, 15 g of Na_2SO_3). Stain the cells in 0.5% $CuSO_4$ for 5 min, followed by a dip in water and 5 min in 1% methyl green. Wash briefly to remove excess dye. For double label in situ hybridization, coat the slides with a thin-barrier film made by dropping 0.4 ml of liquid Krylon (Borden), collected from a spray can, onto a water bath. Add the Krylon within a 40-cm circle of polyethylene tubing. As the solvent evaporates, a thin film of uniform thickness (silver interference color) forms in the center. Remove the tubing, and pick the film up from underneath with a developed and stained slide with magenta grains in the first layer. Dry in a semivertical position. Coat the Krylon film with a second layer of NTB-2 emulsion after 'subbing' the film (0.5% gelatin, 0.05% $CrK(SO_4)_2 \cdot 12\ H_2O$ for 5 min) to provide a substrate for the emulsion to bond to the film. Develop the second layer, and convert the grains to a cyan color.

(2) Flow diagram of procedures to quantitate synthesis of viral nucleic acids in single cells.

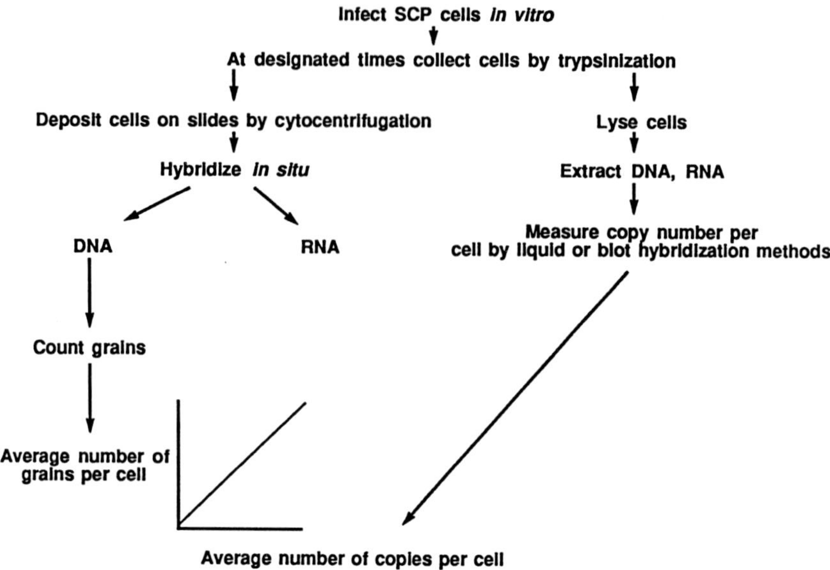

Sheep choroid plexus cell cultures are infected at a mol of 3 PFU/cell. After adsorption of virus for 2 h at 4 °C, infection is initiated synchronously by the addition of warm medium. Cells are collected by trysinization at specified times after infection and the average content of nucleic acid per cell is determined in one aliquot. The remaining cells are sedimented onto slides where, after fixation, the content of nucleic acid in individual cells is assayed by hybridization in situ.

Racz P, Haase AT, Gluckman JC (eds): Modern Pathology of AIDS and Other Retroviral Infections. Basel, Karger, 1990, pp 17–28

In situ Hybridization with Hapten-Modified DNA Probes

A. K. Raap, R. W. Dirks, N. M. Jiwa, P. M. Nederlof, M. van der Ploeg

Department of Cytochemistry and Cytometry, Medical Faculty of Leiden University, Leiden, The Netherlands

The techniques of immunocytochemistry are widely applied for the microscopic identification of cells expressing specific antigens. By linking the immunocytochemical detection methods to the technique of in situ hybridization of nucleic acids, also detection of specific nucleic acid sequences in individual cells has become feasible. This principle has been elegantly shown by Rudkin and Stollar [43] who employed antisera highly specific for DNA-RNA duplexes for the visualization of the rDNA genes in polytene chromosomes of Drosophila after in situ hybridization with rRNA. Following this, Bauman et al. [4] developed a direct in situ hybridization method, in which RNA probes were terminally labeled with a fluorochrome (e.g. FITC or TRITC). This direct method is, however, limited in its sensitivity, and therefore immunocytochemical amplification procedures, using anti-FITC antibodies, have been explored to increase its detection sensitivity [5]. Since then, a number of chemical and enzymic nucleic acid hapten modification procedures has been described which employ this principle of immunocytochemical detection of haptenized nucleic acid probes after in situ hybridization (for a review of methods developed till 1988, see Raap et al. [41]). These methods have several attractive features which make them well suited for application in various disciplines where it is of importance to have molecular knowledge at the DNA and/or RNA level of the individual cell in a morphological context. It is the purpose of this chapter to review the most relevant haptenization procedures for conventional DNA recombinant probes, and to illustrate with a few examples what the limits of current sensitivity of the techniques are and how they are applied in various biomedical fields.

Haptenization of Nucleic Acids

All modifications described below have in common that the presence of the hapten does not affect the hybridization properties of the probes to an appreciable extent at the degrees of modification commonly used (1–10% of all nucleotides).

Biotin

There are several ways of introducing a biotin moiety in DNA or RNA. Currently, the most widely used approach for biotin modification of DNA is the enzymic introduction of biotinated deoxyuridine (either by nick translation or the random priming method). The major achievement for this procedure has been the chemical synthesis of deoxyuridine triphosphate to which biotin is bound – via a linker arm – to the C5 of the uracil base [29]. For RNA, biotinated UTP can be used in a RNA polymerase reaction. As an alternative for enzymic biotination, a photochemical procedure has been described as well as a chemical one using the principle of bisulfite catalyzed transamination described below. Detection of biotinated probes is mediated by antibiotin antibodies or (strept)avidin.

Acetylaminofluorene (AAF)

This haptenization procedure is based on the fact that the carcinogen acetoxy-AAF reacts under mild conditions covalently with the C8 of guanosine residues in DNA or RNA. The resulting AAF modified probe is detected after in situ hybridization via anti-AAF antibodies. DNA as well as RNA, double or single stranded, can be readily labeled with AAF [26, 48].

Digoxigenin

Using similar chemistry for modifying (d)UTP with biotin, Boehringer has introduced the hapten digoxigenin at the C5 position of dUTP. Digoxigenin is incorporated in DNA by nick translation or primer extension.

Mercury Modification

This method distinguishes itself from the previous ones by the fact that the hapten is introduced after the hybridization. First the nucleic acid probe (DNA or RNA, double or single stranded) is chemically mercurated at the C5 position of the pyrimidines. Then the mercurated probe is hybridized, after which it is rendered immunochemically detectable by reacting it with a compound (the ligand) that on the one end carries a sulfhydryl group, which assures firm binding to the probe-bound mercury, and on the other terminus the hapten [18, 19]. The chemical structure of the SH-hapten was found to be of great importance for obtaining good results. Trinitrophenol, biotin and fluorescein have been used as haptens [20].

Sulfonation/Transamination

Bisulfite reacts reversibly with the C5–C6 double bond of cytidine residues in DNA to give 5,6-dihydrocytidine-6-sulfonate. The reversibility of the reaction does not allow the isolation of the modified DNA. However, in a subsequent reaction, nucleophiles like O-methoxyhydroxylamine can be used to convert (transaminate) the cytidines to stable reaction products as in the example of N4-methoxy-5,6-dihydrocytidine-6-sulfonate [47]. As the sulfonation/transamination involves the NH_2 of cytidine involved in Watson-Crick base pairing, there is quite a negative effect on the stability of the modified duplexes. In practice, however, the effect is not of such an order that it influences the specificity of the hybridization reaction, at least for most conventional DNA plasmid probes. There are several ways to use such sulfonated/transaminated probes as 'antigens'. One may elicit antibodies against the sulfonated/transaminated cytidine nucleotide, or one may make use of a hapten in the transaminating compound. The first approach has been used by Orgenics (Israel) who commercialized the procedure. The other approach, using biotin as the hapten, has been described by Viscidi et al. [50] and Reisfeld et al. [42].

The in situ Hybridization Procedure

Nonradioactive in situ hybridization is a fairly complex multistep procedure of which the individual steps have to be optimized against each other, so as to get an optimal balance between morphology and in situ hybridization signal. With limited experimentation and awareness of the potential pitfalls, operational protocols can, however, be designed in a fairly short time for the major applications. In the following some general guidelines will be given. For the major applications detailed protocols can be found in the relevant literature.

Fixation

From the chemical point of view there is little limitation in the type of fixative used as nucleic acids are relatively unreactive to commonly used fixatives, or the reaction is reversible. For instance, in RNA Northern filter hybridizations, the RNA is kept denatured during electrophoresis and blotting by the presence of formaldehyde, but can still be hybridized after removal of formaldehyde. The success of an in situ hybridization, however, is largely determined by such factors as accessibility of reagents and loss of target sequences. A compromise must usually be found between extent of immobilization of cell constituents by fixation and the penetration enhancing and other treatments performed after fixation. For histopathological applications it is of importance to state that in situ hybridization to routinely formalin-fixed, paraffin-embedded material is quite feasible.

Endogenous Enzyme Inactivation

When using peroxidase as the final label, it is usually necessary to inactivate endogenous peroxidase. This is most commonly achieved by incubation in methanol/H_2O_2. Alternatives are mild acid treatments and the glucose/glucose oxidase method of Koller et al. [25]. The latter method may be useful for simultaneous nucleic acid and antigen detection as it is an immunoreactivity-saving method. Endogenous alkaline phosphatase activity is usually inactivated by the in situ hybridization procedure.

Protease Treatments

This is one of the most critical steps in the procedure. Whether, and to what extent, this treatment should be done is dependent on such factors as the type of cells and nucleic acid targets under investigation, the fixation used and therefore the density of the fixed cellular matrix, the type and extent of denaturation used and other (pre)treatments.

Denaturation

To be able to in situ hybridize double-stranded sequences, denaturation of the target is a prerequisite. Acid, alkaline and heat treatments can be used for this purpose. These treatments can, however, affect morphology and again a balance between in situ hybridization signal and morphology must be found experimentally. Concerning in situ DNA hybridization, there tends to be general consensus that heat denaturation, either with or without probe present, yields best results. Also a mild enzymic procedure using exonuclease III for generating single-stranded DNA has been described [49].

In situ Hybridization and Posthybridization Washings

For the factors that influence kinetics of hybridization as well as duplex stability, a wealth of literature is available. Here we like to remark that it is of importance for in situ hybridization to fragment the probe to fragments of 50–500 bp for enhanced probe penetration, and that the labeled probes need not be absolutely unique as the nonrelevant labeled sequences that may hybridize aspecifically can be competed out by proper choice of unlabeled competitor DNA (see also Applications).

Immunological Detection and Light Microscopy

For haptenized probes most often straightforward indirect procedures are used. These days there are several types of immunological markers available for light microscopy: fluorophores, enzymes and colloidal gold. In future one may expect that chemiluminescent and time-resolvable multicolor phosphorescence markers will become available for light microscopical localization studies. The choice as to what type of label, and therefore the type of microscopy, should be used is dependent on several factors. Among these are the spatial resolution required, the presence of autofluorescence, the number of cells to be screened, the need to perform multicolor analysis, the need to perform cytometric quantitation and, more trivial, availability of appropriate microscopes. For problems where high resolution (~0.5 μm) and/or multiple labeling is required, fluorescence microscopy is recommended. Limitations of fluorescence microscopy may be the relatively small field of view. This is related to the fact that maximum fluorescence is obtained with high numerical aperture and therefore high magnification lenses. Bright-field microscopy is recommended when not too high a resolution is required (>0.5 μm), and when large numbers of cells have to be screened for absence or presence of immunocytochemical signal as for example in the detection of rarely occurring virus-infected cells. Also multiple enzyme cytochemical staining procedures are quite feasible. One of the most important features of bright-field microscopy is its relative simplicity and wide availability, making it suitable for routine applications.

Sensitivity

A study in which the various hapten modifications are compared with respect to sensitivity has not yet appeared. However, with all methods it is possible to detect with filter hybridizations sub-picogram quantities of homologous DNA. In practice, the sensitivity will be determined largely by the quality of the immunocytochemical detection system, which includes background suppression measures. As also discussed below, single copy gene detection is possible.

Applications

Virus Detection

Viral diagnosis is an area in which the nonradioactive in situ hybridization procedures have penetrated rapidly. Especially for those viruses that are difficult to diagnose by viral culture or serological techniques, the in situ hybridization methods are welcome additional methods. Prominent examples in histopathology are cytomegalovirus, JC virus, papilloma and Parvo-B19 virus DNA detection [1, 6, 8, 9, 11, 32, 34, 35, 40, 46, 51]. When it comes to histopathological diagnosis of virus infections, the in situ hybridization technique has to compete with other morphological methods, such as conventional cytopathological effect detection in hematoxylin and eosin stained sections and immunocytochemical techniques. The former method is for example for CMV diagnosis inferior to in situ hybridization with respect to sensitivity, as has been reported by several authors. However, when antibodies are available that can give reliable immunocytochemical staining on paraffin sections the immunocytochemical approach may be preferred for routine purposes. For example, in a recent study we described the use of monoclonal anti-CMV antigen antibodies for the immunocytochemical detection after pepsin treatment of routinely formaldehyde-fixed, paraffin-embedded tissue and conclude that in situ hybridization and immunocytochemistry are equally sensitive in detecting CMV in paraffin sections, but that immunocytochemistry is technically less demanding.

Cytogenetics

Also in cytogenetics the methods have found their use rapidly, and it may be anticipated that the in situ hybridization technique will find a place next to the conventional cytogenetic techniques. Clearly, the increased availability of cloned DNA fragments has and will contribute significantly to this development.

The sensitivity of current methods is such that they can be readily applied for probes which recognize repetitive DNA sequences. Thus, in somatic cell hybrids the alien chromosomes (segments) are easily detected using Kpn- or Alu-like repeat sequences, or more simply, total DNA as probe [16, 33, 44]. With probes that recognize clustered repetitive sequences, more specific detection of chromosomes (segments) is possible. Particularly, the satellite DNA recombinant probes have been useful in this respect. They allow the rapid assessment of the chromosome (segment) copy number, not only in metaphase but also in interphase nuclei [10, 12,

14]. Next to application in early, prenatal diagnosis [23], this methodology is of considerable importance in (solid) tumor cytogenetic studies as it is often cumbersome to obtain sufficient, representative and high-quality metaphase spreads for banding procedures [21, 37]. Cremer et al. [12] have proposed the term 'interphase cytogenetics' for this type of work. In this type of research the double hybridization procedures are of great value as they provide more information than the sum of the single hybridization [18, 19, 37]. With the introduction of the blue immunofluorophore AMCA [24], triple fluorescence in situ hybridization with chromosome-specific repetitive probes proved possible [36]. Recently, we have also performed quadruple simultaneous hybridizations. For this purpose the fourth probe was labeled with two haptens (e.g. biotin and AAF), and was identified on the basis of its double fluorescent color [38].

For single copy gene localization an extreme sensitivity is required as it demands the detection of one DNA sequence weighing only a few attograms. With current technology this is feasible, as illustrated by the work of several groups [2, 3, 7, 17, 27, 30]. It is of considerable importance to note that for in situ hybridization with cloned genomic sequences, the labeled cloned DNA probe needs not to be free of repetitive sequences for specific gene(s) localization. The repeat DNA (e.g. Alu repeats) which can be present in genomic clones can effectively be eliminated from participating in the in situ hybridization by competition hybridization with unlabeled total human DNA [28]. This allows, for instance, cosmid clones to be fine mapped, as well as recombinant libraries from flow-sorted chromosomes to be used in so-called 'chromosome painting' [13, 31, 39].

mRNA Detection

When compared with DNA detection, the number of articles describing nonradioactive mRNA detection is by far less. This reflects probably the special problem that can be associated with mRNA detection in situ. In our study we chose to use as a biological model for developing reliable methods, a system in which specific mRNAs are expressed in high copy number. It concerns here the neuropeptidergic system of *Lymnaea stagnalis,* specifically the egg-laying hormone (ELH) and a molluscan insulin-like peptide (MIP), produced in the polyploid caudodorsal cells and in the light green cells, respectively [15]. As with DNA detection, an optimal balance between RNA retention (which is closely associated with morphology) and in situ hybridization signal must be found. Though with almost any type of fixation in situ hybridization signals can be obtained, it was found that a freeze dry, paraformaldehyde vapor fixation in combination with a pepsin pretreatment gave optimal results with respect to signal and morphology. A striking difference with respect to the effect of protease pretreatment was found between the two neuropeptide hormones: with the ELH mRNA, signals can be obtained without pretreatment and with MIP mRNA only after such a treatment. Recently we have obtained excellent results with directly FITC-labeled synthetic oligonucleotides for these hormone mRNAs [Dirks et al., submitted]. The attractiveness of such oligonucleotides as hybridization probes resides in the fact that they are well-defined, small molecules which penetrate the tissue sections quite well. This may actually compensate to some extent for the smaller target they recognize when compared with cDNA probes.

Hybridization and Immunocytochemistry versus the Polymerase Chain Reaction

One of the favorable features of identification of virus-infected cells by microscopic cytochemical techniques is its sensitivity. In principle, one infected cell amongst hundreds of thousands of uninfected cells can be detected. Conventional Southern or dot-blot filter hybridization techniques are too insensitive to detect viral DNA in such a situation. In addition, these techniques will not give morphological information on the identity of the infected cell types. Recently, however, a new DNA analysis technique has been introduced which has extreme sensitivity. It concerns here the polymerase chain reaction (PCR), which is based on the exponential, in vitro amplification of target sequences by repeated cycles of denaturation, primer annealing and polymerization using a heat-stable DNA polymerase [45]. This amplification greatly facilitates further analysis and detection. The sensitivity is such that one target sequence in a sample can be detected. PCR, therefore, is a strong competitor of the microscopic techniques with respect to detection of e.g. viral DNA.

In a recent study we have compared PCR and immunocytochemistry for the detection of cytomegalovirus viremia. It was found that both techniques are quite well capable of detecting virus-infected cells occurring in low frequency (1:100,000 to 1:1,000,000 [Jiwa et al., in preparation]). Thus, it appears that the sensitivity of both techniques is determined by the number of cells that can be accommodated in an assay. The PCR, however, remains a nonmorphological technique.

Concluding Remarks

The techniques of nonradioactive in situ hybridization have matured in recent years to procedures that for a number of important biomedical problems can be readily applied. Their great virtues are sensitivity and specificity, but perhaps the most important one is the capability of bridging the gap that exists between the morphological and molecular (bio)medical disciplines.

References

1 Aksamit AJ, Mourrain P, Sever JL, et al: Progressive multifocal leukoencephalopathy: Investigation of three cases using in situ hybridization with JC virus biotinylated DNA probes. Ann Neurol 1985;18:490–496.
2 Albertson DG: Mapping muscle protein genes by in situ hybridization using biotin-labeled probes. EMBO J 1985;4:2493–2498.
3 Ambros PF, Matzke MA, Matzke AJM, et al: Detection of a 17 kb unique (T-DNA) in plant chromosomes by in situ hybridization. Chromosoma 1986;94:11–16.
4 Bauman JGJ, Wiegant J, Borst P, et al: A new method for fluorescence microscopical localization of specific DNA sequences by in situ hybridization of fluorochrome-labeled RNA. Exp Cell Res 1980;138:485–490.
5 Bauman JGJ, Wiegant J, Van Duijn P: Cytochemical hybridization with fluorochrome-labelled RNA. III. Increased sensitivity by the use of anti-fluorescein antibodies. Histochemistry 1981;73:181–193.
6 Beckmann AM, Myerson D, Daling JR, et al: Detection and localization of human papillomavirus in human genital condylomas by in situ hybridization with biotinylated probes. J Med Virol 1985;16:265–273.
7 Bhatt B, Burns J, Flannery D, et al: Direct visualization of single copy genes on banded chromosomes by nonisotopic in situ hybridization. Nucleic Acids Res 1988;16:3951–3961.
8 Boerman RH, Arnoldus EPJ, Raap AK, et al: Diagnosis of progressive multifocal leukoencephalopathy by hybridization techniques. J Clin Pathol 1989;42:153–161.
9 Brigati DJ, Myerson D, Leary JJ, et al: Detection of viral genomes in cultured cells and paraffin-embedded tissue sections using biotin-labeled hybridization probes. Virology 1982;126:32–50.
10 Burns J, Chan VTW, Jonasson JA, et al: Sensitive system for visualizing biotinylated DNA probes hybridized in situ: rapid sex determination of intact cells. J Clin Pathol 1985;38:1085–1092.
11 Burns J, Redfern DRM, Esiri MM, et al: Human and viral gene detection in routine paraffin-embedded tissue by in situ hybridization with biotinylated probes: viral localisation in herpes encephalitis. J Clin Pathol 1986;39:1066–1073.
12 Cremer T, Landegent J, Brueckner A, et al: Detection of chromsome aberrations in the human interphase nucleus by visualization of specific target DNAs with radioactive and non-radioactive in situ hybridization techniques: diagnosis of trisomy-18 with probe L1.84. Hum Genet 1986;74:346–352.

13 Cremer T, Lichter P, Borden J, et al: Detection of chromsome aberrations in metaphase and interphase tumor cells by in situ hybridization using chromosome specific library probes. Hum Genet 1988;80:235–246.

14 Devilee P, Thierry F, Kievits T, et al: Detection of chromosome aneuploidy in interphase nuclei from human primary breast tumors using chromosome specific repetitive DNA probes. Cancer Res 1988;48:5825–5830.

15 Dirks RW, Raap AK, Van Minnen J, et al: Detection of mRNA molecules coding for neuropeptide hormones of the pond snail *Lymnaea stagnalis* by radioactive and non-radioactive in situ hybridization: a model study for mRNA detection. J Histochem Cytochem 1989:37:7–14.

16 Durnam DM, Gelinas RE, Myerson D: Detection of species chromosomes in somatic cell hybrids. Somat Cell Mol Genet 1985;11:571–577.

17 Garson JA, Van den Berghe JA, Kemshead JT: Novel non-isotopic in situ hybridization technique detects small (1 kb) unique sequences ion routinely G-banded human chromosomes: fine mapping of N-myc and b-NGF genes. Nucleic Acids Res 1987;15:4761–4770.

18 Hopman AHN, Wiegant J, Raap AK, et al: Bi-color detection of two target DNAs by non-radioactive in situ hybridization. Histochemistry 1986;85:1–4.

19 Hopman AHN, Wiegant J, Tesser GI, et al: A non-radioactive in situ hybridization method based on mercurated nucleic acid probes and sulfhydryl-hapten ligands. Nucleic Acids Res 1986;14:6471–6488.

20 Hopman AHN, Wiegant J, Van Duijn P: Mercurated nucleic acid probes, a new principle for non-radioactive in situ hybridization. Exp Cell Res 1987;169:357–368.

21 Hopman AHN, Ramaekers FCS, Raap AK, et al: In situ hybridization as a tool to study numerical chromosome aberrations in solid bladder tumors. Histochemistry 1988;89:307–316.

22 Jiwa NM, Raap AK, Van de Rijke FM, et al: Detection of human cytomegalovirus DNA antigens in formaldehyde-fixed material from bone marrow transplant patients. J Clin Pathol 1989;42:749–754.

23 Julien C, Bazin A, Guyot B, et al: Rapid prenatal diagnosis of Down's syndrome with in situ hybridization of fluorescent DNA probes. Lancet 1986;ii:863–864.

24 Khalfan H, Abuknesha R, Rand-Weaver R, et al: Aminomethyl coumarin acetic acid: a new fluorescent labeling agent for proteins. Histochem J 1986;18:497–499.

25 Koller U, Stockinger H, Majdic O, et al: A rapid and simple immunoperoxidase staining procedure for blood and bone marrow samples. J Immunol Methods 1986;86:75–81.

26 Landegent JE, Jansen in de Wal N, Baan RA, et al: 2-Acetylaminofluorene-modified probes for the indirect hybridocytochemical detection of specific nucleic acid sequences. Exp Cell Res 1984;153:61–72.

27 Landegent JE, Jansen in de Wal N, Ommen GJB, et al: Chromosomal localization of a unique gene by non-autoradiographic in situ hybridization. Nature 1985;317:175–177.

28 Landegent JE, Jansen in de Wal N, Dirks RW, et al: Use of whole cosmid cloned genomic sequences for chromosomal localization by non-radioactive in situ hybridization. Hum Genet 1987;77:366–370.

29 Langer PR, Waldrop AA, Ward DA: Enzymatic synthesis of biotin-labeled polynucleotides: novel nucleic acid affinity probes. Proc Natl Acad Sci USA 1981;78:6633–6637.

30 Lawrence JB, Villnave CA, Singer RH: Sensitive high resolution chromatin and chromosome mapping in situ: presence and orientation of two closely integrated copies of EBV in a lymphoma line. Cell 1988;52:51–61.

31 Lichter P, Cremer T, Borden J, et al: Delineation of individual human chromosomes in metaphase and interphase cells by in situ hybridization using recombinant DNA libraries. Hum Genet 1988;80:224–234.
52 Loning T, Milde K, Foss HD: In situ hybridization for the detection of cytomegalovirus (CMV) infection. Application of biotinylated CMV-DNA probes on paraffin-embedded specimen. Virchows Arch 1986;409:777–790.
33 Mitchell AR, Ambros P, Gosden JR, et al: Gene mapping and physical arrangements of human chromatin in transformed, hybrid cells fluorescent and autoradiographic in situ hybridization compared. Somat Cell Mol Genet 1986;12:313–324.
34 Myerson D, Hackman RC, Nelson JA, et al: Widespread occurrence of histologically occult cytomegalovirus. Hum Pathol 1984;15:430–439.
35 Myerson D, Hackman RC, Meyers JD: Diagnosis of cytomegalovirus pneumonia by in situ hybridization. J Infect Dis 1984;150:272–277.
36 Nederlof PM, Robinson D, Wiegant J, et al: Three color fluorescence for the simultaneous detection of multiple DNA targets. Cytometry 1989;10:20–27.
37 Nederlof PM, Van der Flier S, Raap AK, et al: Detection of chromosome aberrations in interphase nuclei by in situ hybridization. Cancer Genet Cytogenet, in press.
38 Nederlof PM, Van der Flier S, Wiegant J, et al: Multiple fluorescence in situ hybridization. Cytometry, in press.
39 Pinkel D, Landegent J, Collins C, et al: Fluorescence in situ hybridization with human chromosome specific libraries: detection of trisomy 21 and translocation of chromosome 4. Proc Natl Acad Sci USA 1988;85:9138–9142.
40 Raap AK, Geelen JL, Van der Meer JWM, et al: Non-radioactive in situ hybridization for the detection of cytomegalovirus infections. Histochemistry 1988;88:367–373.
41 Raap AK, Hopman AHN, Van der Ploeg M: Use of hapten modified nucleic acid probes in DNA in situ hybridization; in Bullock G, Petrusz P (eds): Techniques in Immunocytochemistry. New York, Academic Press, 1989, vol IV.
42 Reisfeld A, Rothenberg JM, Bayer EA, et al: Nonradioactive hybridization probes prepared by the reaction of biotin hydrazide with DNA. Biochem Biophys Res Commun 1987;142:519–526.
43 Rudkin GT, Stollar BD: High resolution detection of DNA-RNA hybrids in situ by indirect immunofluorescence. Nature 1977;265:472–473.
44 Schardin M, Cremer T, Hager HD, et al: Specific staining of human chromosomes in Chinese hamster × man hybrid cell lines demonstrates interphase chromosomes territories. Hum Genet 1985;71:281–287.
45 Saiki RK, Gelfland DH, Stoffel S, et al: Primer directed enzymatic amplification of DNA with a thermostable DNA polymerase. Science 1988;239:487–491.
46 Salimans M, Van der Rijke FM, Raap AK: Detection of Parvo-B19 DNA in fetal tissue by in situ hybridization and polymerase chain reaction. J Clin Pathol 1989;42:525–530.
47 Sverdlov ED, Monastyrskaya GS, Guskova LI, et al: Modification of cytidine residues with a bisulphite-O-methylhydroxylamine mixture. Biochim Biophys Acta 1974; 340:153–165.
48 Tchen P, Fuchs RPP, Sage E, et al: Chemically modified nucleic acids as immunodetectable probes in hybridization experiments. Proc Natl Acad Sci USA 1984;81: 3466–3470.
49 Van Dekken H, Pinkel D, Mullikin J, et al: Enzymatic production of single-stranded DNA as a target for fluorescence in situ hybridization. Chromosoma, 1988;97:1–5.

50 Viscidi RP, Connelly CJ, Yolken RH: Novel chemical method for the preparation of nucleic acids for non-isotopic hybridization. J Clin Microbiol 1986;23:311–317.

51 Walboomers JMM, Mullink H, Melchers WJG, et al: Sensitivity of in situ detection with biotinylated probes of HPV 16 DNA in frozen tissue sections of squamous cell carcinoma of the cervix. Am J Pathol 1988;131:587–594s.

A. K. Raap, Department of Cytochemistry and Cytometry, Medical Faculty of Leiden University, Wassenaarsweg 72, NL–2333 AL Leiden (The Netherlands)

Racz P, Haase AT, Gluckman JC (eds): Modern Pathology of AIDS and Other Retroviral Infections. Basel, Karger, 1990, pp 29–41

Diagnostic Applications of Nonisotopic in situ Hybridization for HIV or EBV Nucleic Acids

R. Singer[a], *J. B. Lawrence*[a], *R. Bashir*[c], *K. Byron*[b], *J. Sullivan*[b]

Departments of [a]Cell Biology and [b]Pediatrics, University of Massachusetts Medical School, Worcester, Mass.; [c]Department of Neurology, University of Nebraska Medical Center, Omaha, Nebr., USA

Modern technology has provided tools helpful to the clinical researcher which can precisely identify the presence and type of a particular etiologic agent, directly within cells or tissues. Immunocytochemistry has been directed primarily toward detecting antigens expressed by viral infection. In situ hybridization detects the nucleic acid sequences of the virus, and as such, does not rely on expression of viral antigens or even transcription of viral RNA. Furthermore, nucleic acid hybridization is not affected by fixatives used during tissue or cell processing as often are the epitopes detected by many antibodies. Over the past several years, in situ hybridization has been used as a diagnostic tool to indicate viral infection [for example, see 3, 4, 7, 11, 26, 31]. Most work has used isotopically labelled DNA or RNA probes and the hybridization detected with autoradiography. This method is cumbersome because of the need to coat slides with emulsion in the dark, expose and develop them. The resultant silver grains are indicative of hybridization to particular cells, but are visually difficult to identify clearly against a stained background without the help of auxillary methods of detection such as dark field microscopy. Furthermore, the length of the track of the decay particle limits the cellular resolution and hence is inferior to immunofluorescence.

Recent work has emphasized the use of nonisotopic detection methods to improve the convenience and speed of the procedure [5, 17, 27, 29, 32, 33]. Nonisotopic means of detection of viral nucleic acids have been a subject of intense investigation since the report of Langer et al. [14, 15]. The increase in resolution afforded by the nonisotopic methods, particularly that of fluorescence, has allowed significant increases in sensitivity so that a single copy of a virus is detectable in a single cell [18]. This is possible because the probe is

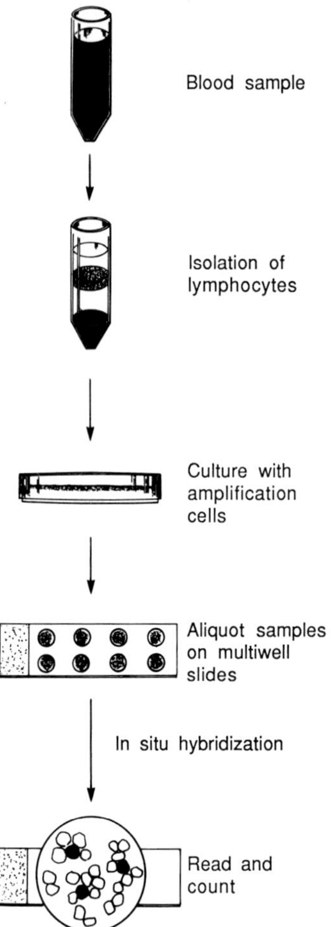

Fig. 1. Schematic diagram illustrating the preparation of cells for in situ hybridization.

detected at the site of hybridization. The elimination of autoradiography has therefore made this method more convenient and rapid. In addition, the ease of visualization of a labelled cell within a population of hundreds of thousands of negative cells has been made relatively simple using this approach [27].

There are two types of clinical situations for which in situ hybridization in general and nonisotopic detection in particular are important. The first is where the infected cell is extremely rare in a population of uninfected cells, such that molecular techniques currently in use are not sufficiently sensitive to overcome the dilution effect of the nucleic acids from the uninfected cells. Such a case is exemplified by examination of peripheral blood lymphocytes from individuals seropositive for HIV [9]. These individuals have been shown to have a level of infected cells below 10^{-4} [11]. The second case for improved in situ hybridization technology is where a latent virus exists in infected tissue, and may be the cause of neoplastic transformation, but very low levels of viral proteins and/or nucleic acids are made. Such a case is exemplified by the EBV-transformed lymphomas. Namalwa cells, for instance, have been isolated from an African Burkitt's lymphoma and have two copies of EBV integrated into the host genomic DNA [12, 18]. This and other cell lines resulting form herpes virus latency provide a model for tissues resulting from similar events, such as lymphomas of the central nervous system.

We have used both EBV and HIV as models for two types of viruses: double-stranded DNA and single-stranded RNA respectively. In addition, EBV has been investigated in solid tumor tissue as well as in transformed lymphoblastoid cells [2]. We describe here methodology specifically related to the detection of HIV in a clinical setting. The details of this work are contained in a previous publication [27].

Methods

We have devised methods for detection of small numbers of HIV-infected cells either directly from patients or from coculture of patient lymphocytes with mitogenically stimulated normal cells. The outline of the approach is schematized in figure 1.

Normal lymphocytes were obtained from individuals after venipuncture and Hypaque-Ficoll separation, and deposited on 8-well slides (cell line) at high concentration (e.g., 3×10^7/ml). The residual cells were removed by micropipette such that each well contained approximately 50,000 cells. Cells were then fixed in 4% paraformaldehyde for 10 min and stored in 70% ethanol. Slides are rehydrated in a phosphate-buffered saline, and then in 0.1 *M* Tris–0.2 *M* glycine, and finally in 50% formamide, 2 × SSC. Maximal hybridization occurs with high probe concentration; 40 ng of probe DNA is generally used. Biotin-11-dUTP from Enzo Biochemicals, biotin-16-dUTP from Boehringer Mannheim or biotin-14-dATP from Bethesda Research Laboratories all are sufficient to label the probe DNA using nick translation. A cloned segment of HIV-1 (IIIB) obtained through NIH containing 8.9 kb of the viral genome was used as the probe [22, 25]. Nick translation conditions were as follows: 1.5 μl of 400 μ*M* biotin dUTP (BRL), 1 μl of dACG mix (dATP, dCTP, dGTP 600 μ*M* each; PL Biochemicals) 1 μl of 10 × nick translation buffer (0.5 *M* Tris Cl, pH 7.2; 0.1 m*M* $MgSO_4$; 1 m*M* dithiothreitol) 500 μg/ml bovine serum albumin; BSA Pentax Frac V), 5 μl sterile glass distilled H_2O; 1 μl of 0.1 μg/μl HIV DNA, 1 μl of DNase (final 34 ng/ml: concentration determines the probe size; large probe sizes cause background), 1 μl of DNA Polymerase I (Boehringer). Incubation is 3 h at 15°C, then 90 μl of 50 m*M* EDTA and 1 μl of 10% SDS is

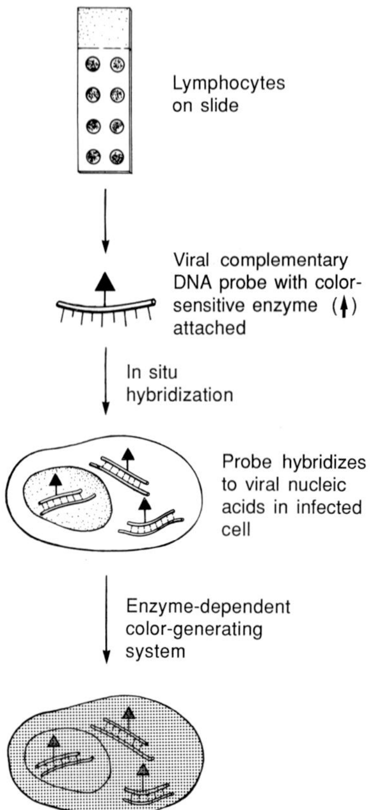

Fig. 2. Schematic diagram illustrating the use of nonisotopically labelled probe. In this case, the probe is directly labelled by enzme (see fig. 3f), but in most of the work presented here, the triangle represents a biotin moiety detected by a streptavidin-alkaline phosphatase conjugate with identical color generation (see fig. 3a–e).

added. Purification from free nucleotides is with a sterile Sephadex G-50 'spin' column packed in a 1-ml disposable syringe. Nick translated probe is aliquoted into an Eppendorf tube and lyophilized with carrier nucleic acids (final concentration of 100 μg/ml of sheared salmon sperm and tRNA). Deionized formamide is mixed with the lyophilite and placed in a 90°C heating block for 10 min. Hybridization buffer is prepared by mixing 20 × SSC; 10% BSA, dextran sulfate (50% solution, autoclaved in water) and water in equal quantities. Slides are then quickly utilized after blotting away excess solution with the deposition of the heated probe mixed rapidly with an equal volume of hybridization buffer onto each serological well, covered with a small strip of parafilm, placed for 3 h (or overnight as is convenient) in a humidified 37°C incubator. Slides are rinsed in 50% formamide/2 × SSC for 30 min at 37°C, then placed in 2 × SSC for 30 min and finally in 1 × SSC for 30 min (or overnight).

A streptavidin-alkaline phosphatase conjugate (Dakopatts) was used with a dilution 1:250 into 4 × SSC with 1% fatty acid free BSA and exposed to the hybridized cells for 20 min. After washing in 4 ×SSC, the cells were put into a pH 9.5 wash (0.1 *M* Tris, 0.1 *M* NaCl, 50 m*M* $MgCl_2$) previous to color development with NBT and BCIP (developed 10–20 min) after Singer et al. [29]. Oligonucleotide probes were obtained from Dupont Corp. ('SNAP' probes) and Molecular Biosystems, Inc. A mixture of 14 oligonucleotides conjugated directly to alkaline phosphatase and synthesized to invariant HIV sequences were used. The probe concentration of the final solution was 6 n*M*. Hybridization was in 5 × SSC, 0.5% SDS and 1% BSA at 50°C for 20 min. Samples were washed at 42°C for 5 min in 1 × SSC. Alkaline phosphatase detection was with nitroblue tetrazolium and bromochloroindolyl phosphate from 6 h to overnight. A schematic of this approach is presented in figure 2 and the result in figure 3f.

Nonisotopic Detection of HIV Nucleic Acids in situ

Three kinds of cell samples were investigated using this approach. The first was a culture of normal human lymphocytes stimulated mitogenically with PHA and IL-2 for 3 days and then infected with HIV for 7 days; the second was a sample of patient lymphocytes cocultured with normal human lymphocytes for 21 days and the third was a peripheral blood lymphocyte sample taken directly from a seropositive individual. Control samples consisted of uninfected lymphocytes in culture or directly from normal individuals. Illustrations of the results achieved using this approach are seen in figure 3.

The culture infected in vitro for 7 days showed a large number of positive cells (fig. 3a) which varied from 12 to 27% depending on the multiplicity of infection as well as conditions. The infected cells were intensely colored with the alkaline phosphatase reaction. While there was considerable variability in the intensity of color from cell to cell, presumably reflecting the amount of viral nucleic acids per cell [30], the negative cells were easily distinguishable from the positive. When this approach was applied to an uninfected cell culture, no cells were scored positive when more then 10^5 cells were microscopically analyzed. Therefore, the false-positive rate is less than 10^{-5}, and the in situ hybridization approach is, at the least, sufficiently accurate to score infected cells in the 10^{-4} range. It is import to emphasize that positive cells have never been seen in negative control samples. This means that even a single positive cell is a significant event.

The same procedure was applied to the detection of positive cells when normal lymphocytes were cocultured with cells from seropositive but asymptomatic individuals (fig. 3b–d). The patients used in this study were from the hemophiliac population. In addition, supernatants from these cultures were

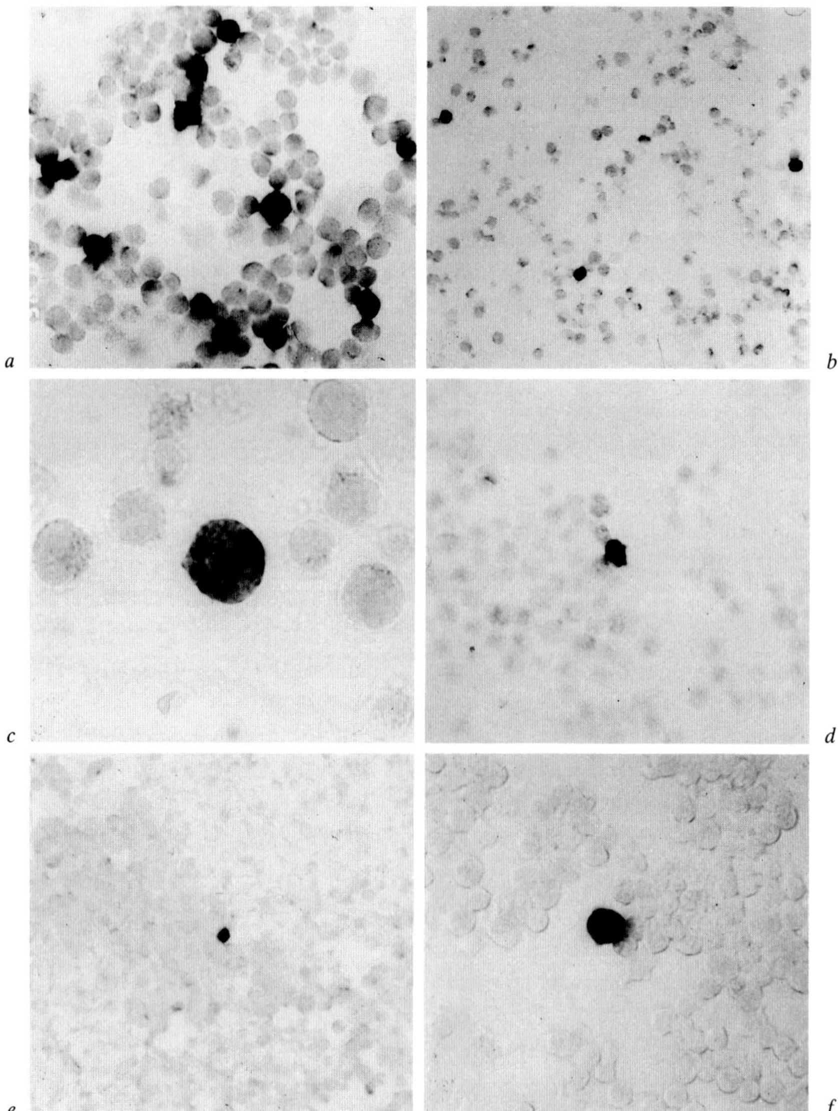

Fig. 3. Various detections of rare HIV infection in lymphocytes using nonisotopic in situ hybridization. *a* Normal human peripheral blood lymphocytes stimulated with PHA and IL-2 and grown in culture for 7 days after infection with HIV at low MOI, positive cells are at the 10^{-1} level. *b* Coculture of normal PBL with patient lymphocytes (3:1) for 21 days. A level of 10^{-3} positive cells is detectable in this culture. *c* A syncytial cell seen in a coculture. *d* Coculture of normal PBL with patient lymphocytes. A level of 10^{-4} positive cells is detected in this culture.

assayed for the viral core antigen, P24, using the first-generation Dupont antigen-capture test. Aliquots were taken from the cultures either for in situ hybridization or for P24 assay at four intervals over a 21-day period. An analysis of the cultures obtained from 30 seropositive individuals and 2 seronegative individuals revealed that in situ hybridization detected positive cells at the 10^{-4} to 10^{-3} level. Cultures were defined as virus positive more often by in situ hybridization (55%) than by the antigen-capture test (27%). Neither of 2 seronegative individuals had any positive cells detected in the culture (10^5 cells analyzed by microscopy). Also of interest was the observation that in situ hybridization could be predictive of the presence of virus, since it often detected positive cells before the supernatant was found positive for viral antigen, when antigen was present at all. Occasionally however, in situ results which were positive early during the culture became negative subsequently, and positive once again. This phenomenon occurs frequently enough that it may be indicative of some aspect of the infection, or of the technique. A possible explanation involving the infection could be that the virus is replicating in lytic cycles. A methodological explanation is that few cells are positive (10^{-4} to 10^{-5}) so that the sample size of 50–100,000 cells is too small and analysis of more cells is necessary for a statistically significant result. In support of this possibility is the observation that most positive results are based on 2–3 positive cells per well (approximately 50,000 cells). Sampling variations resulting in false-negative determinations using 10^5 cells have also been seen using the polymerase chain reaction (PCR), confirming the above observations [13, 23].

In situ hybridization is an effective method for detecting and quantitating HIV-positive cells in culture. Used in conjunction with the P24 assay, it allows a quantitative verification, and may serve as a diagnostic tool in the absence of other tests. A particularly useful application of the in situ approach may be as a rapid and accurate means for screening experimental drugs. The positive cells serve as a miniplaque assay to determine concentration effects, or even to test a patient's cells responsiveness to a particular drug.

---

e Patient lymphocytes were deposited on a slide after Ficoll-Hypaque separation and positive cells detected directly without coculture. The positive cell represents between the 10^{-4} and 10^{-5} level of detection. *f* A patient coculture as in figure 3c, except detected with the mixture of three oligonucleotide probes ('SNAP' probe, Dupont, MBI) and developed overnight in the alkaline phosphatase color reaction. Cells were photographed in phase so that negative cells are visible. All lymphocytes are 7 μm in diameter.

The in situ hybridization technique, because of its effectiveness in detecting the rare, infected cell, can be used directly on patient samples where the positive cells are reduced 10-fold in concentration from those found in culture (fig. 3e). Despite the decrease in detectable positive cells, lymphocytes from seropositive, asymptomatic hemophiliacs, when analyzed, were found to contain enough virus-infected cells so that approximately 33% of the individuals were immediately scored as virus-positive. This can be compared to the 55% found to be virus-positive when using in situ hybridization to assay cultures containing the patient's cells. Peripheral blood lymphocytes taken directly from seropositive individuals and assayed for virus-positive cells by in situ hybridization not only avoids the processing by tissue culture in order to detect the viral sequences, but it may be useful prognostically to know the number of virus-positive cells circulating in the patient. This also allows analysis of the infection as it exists in vivo rather than in artificial culturing systems removed form immune surveillance.

Although the asymptomatic hemophiliac population was detected to be 55% positive, improvements in culturing as well as in detecting positive cells directly from these individuals will increase the number of positive individuals still further. Recent reports suggest that all of the hemophiliac population which is seropositive may be also virus-positive [6]. It should be clarified that development of this nonisotopic in situ methodology for detecting the more positive cell is an ongoing process. Further improvements are focusing on increased sensitivity capable of detecting a single copy of integrated HIV provirus [Lawrence et al., submitted].

Nonisotopic Detection of DNA Viruses

The detection of a DNA virus would be expected to require a modification in methodology from the detection of HIV since double-stranded target molecules must be denatured in order to hybridize to the probe. In addition, most of these sequences reside in the nucleus (RNA as well as DNA) so that probe must penetrate well within the viscous nucleus in order to hybridize, and be removed if it does not. We have used as a model system an EBV transformed cell line where a single copy of EBV has been integrated into the host chromosome 1 DNA.

We have shown in previous work [18] that this single copy of EBV can be detected in a single cell. The cell line used for this work was Namalwa, originally isolated from an African Burkitt's lymphoma. Metaphase chromosome spreads of these cells revealed that as little as 2 kb of EBV genomic DNA could be detected on each of the sister chromatids with the high resolution technique using fluorescein avidin.

Further work not done on chromosomes but on interphase nuclei from latent or productive cells has shown that the signal from viral RNA is very strong relative to the DNA – perhaps 50 times stronger [17]. The intensity of RNA signal in the latent cells affords the opportunity to improve the sensitivity of detection of viral nucleic acids. Work done on latent cells such as Raji, IB-4, or patient transformed lines showed considerable RNA signal, in many cases multifold within the nucleus compared to one or two foci in Namalwa cells. Ramos cells (EBV-negative) showed no signal. The productively infected cell line in marmoset cells, B95-8, showed many cells with considerably strong signal (about 30%) and some smaller percent (5–10%) extremely bright in fluorescence. The latter would correspond with the small number of lytically infected cells, as has been reported previously [21].

The same cells have been used to ascertain the usefulness of enzymatic detection of the biotinated probe after the method of Leary et al. [19], Unger et al. [33] or Singer et al. [29]. While it is immediately evident that fluorescent detection, due to its superior resolution, has a greater sensitivity, there are some advantages to enzymatic detection. One is that the enzymatic detection is a permanent record, with visual results similar to that obtained by common histochemical stains. Contrarily, fluorescence fades with time. Another is that specialized optical equipment, such as epifluorescence optics, is not necessary. Finally, it circumvents the problem, more common in some tissues than others, of autofluorescence which increases background and therefore decreases signal-to-noise ratios. Since much of in situ analysis is directed toward identifying positive cells, the higher degree of subcellular resolution resulting from the use of fluorescence may not be required. Alkaline phosphatase detection proved adequate to detect even the latently infected cells, Namalwa, although the detection relied on the RNA transcripts for signal rather than the single copy of DNA. Since it would be expected that any cell infected with EBV would be expressing genes of that virus, the detection of RNA would appear to be a viable diagnostic approach. While this work in detail has been submitted elsewhere [1, 2], some of the conclusions are briefly described below.

In order to evaluate the use of clinical tissue with these methods, the cells described above were pelleted in serum, fixed and processed as if they were biopsy tissue in a pathology laboratory. Paraffin sections were then taken and in situ hybridization performed for detection of EBV. Processing the cells through embedment and sectioning significantly reduced signal. Most of this reduction was due to the fact that the signal in each individual cell was not necessarily contained within the section. It may be in addition that penetration of the reagents into this tissue model is also limited. The 'tissue' sections have approximately 35% of the cells detected as positive, even though the same cells when cytospun onto coverslips were detected as almost all positive. Hence, in routine tissue sections, only about a third of the cells will indicate EBV nuclear signal.

The background of alkaline phosphatase staining is occasionally increased in tissue sections, indicative perhaps of increased difficulty in clearing the tissue of reagents. For tissue, simultaneous denaturation of probe and target were performed in order to detect viral DNA as well as RNA. This procedure was first used by Brigati et al. [5] on CMV. Since it appears that most of the signal may be RNA, this step may not be necessary, and gentler methods may yield greater sensitivity. If, however, penetration of the probe into the tissue takes sufficient time that competitive reannealing of probe or target causes signal reduction, then the denaturing step may be useful. Current work is elucidating the requirements for detection in tissue sections. In order to evaluate the applicability of the above techniques, material taken from biopsy on immunocompromised CNS lymphoma patients was analyzed by in situ hybridization. Four patients analyzed were found to be EBV-positive and 4 nonimmunocompromised patients were EBV-negative [1]. One of these immunocompromised patients was confirmed positive by Southern blot techniques. However, in situ hybridization was more sensitive than the blots as well as being more rapid and informative as to tissue type infected.

Value of in situ Diagnostics

We have described here some of the results of methodology used to detect viral RNA and DNA in situ. The value of this approach is evident in the nature of the results: the nucleic acids are detected in the cells irrespective of their expression into proteins. Because every cell can be analyzed independently, and the technique requires little sacrifice of cell morphology, this approach is compatible with standard pathological analysis using microscopy. As such, it is a new and powerful tool in the arsenal of pathologists, hematologists or microbiologists, but conforms to standard histochemical protocols rather than requiring new, expensive equipment or highly skilled personnel.

The protocols of this work can be formulated into a kit, wherein reagents are quality controlled, and positive and negative control samples can aid in confidence of data interpretation. A highly reproducible and sensitive in situ approach would be a useful adjunct to antibody-based cellular diagnostics because of the antigenicity reduction in fixed tissue, and the stability of reagents. Furthermore, the quantitative information obtained through use of the kit would be useful for therapy evaluation centers where actual infected cells and viral load may be important indicators, for instance in infected but seronegative individuals [8, 20, 34]. Finally, pediatric samples [24] can be evaluated for the presence of viral nucleic acids where maternal antibodies may interfere. By making in situ hybridization independent of laboratory training, the future role of in situ diagnostics will be greatly enhanced.

Acknowledgements

We would like to acknowledge support from NIH HD18066 and a contract for the detection of HIV nucleic acids, HB67027. We appreciate the technical assistance of John McNeil, Jerry McNeil, Carol Villnave and Lisa Marselle. We thank Tom Sharpe, Barbara Young and Karl Adler at Dupont Corporation, Billerica, Mass. and Jerry Ruth at Molecular Biosystems, San Diego, Calif. for the gift of the oligonucleotide probes.

References

1 Bashir, R.; Harris, N.; Hochberg, F.; Singer, R.: Detection of EB virus in CNS lymphomas by in situ hybridization. Neurology *39:* 813–817 (1989).
2 Bashir, R.; Hochberg, F.; Singer, R. H.: Detection of Epstein-Barr virus nucleic acids by in situ hybridization: progress toward development of a non-isotopic diagnostic test Am. J. Pathol. *135:* (in press, 1989).
3 Biberfeld, P.; Chayt, K.; Marselle, L. M.; Biberfeld, G.; Gallo, R. C.; Harper, M. E.: HTLV-III expression in infected lymph nodes and relevance to pathogenesis of lymphadenopathy. Am. J. Pathol. *125:* 436–442 (1986).
4 Brahic, M.; Haase, A. T.: Detection of viral sequences of low reiteration frequency by in situ hybridization. Proc. Natl. Acad. Sci. USA *75:* 6125–6129 (1978).
5 Brigati, D. J.; Myerson, D.; Leary, J. J.; et al.: Detection of viral genomes in cultured cells and paraffin-embedded tissue sections using biotin-labelled hybridization probes. Virology *126:* 32–50 (1983).
6 Jackson, J.; Sannerud, K. J.; Hopsicker, J. S.; Kwok, S. Y.; Edson, J. R.; Balfour, H. H.: Hemophiliacs with HIV antibody are actively infected. JAMA *260:* 2236 (1988).
7 Busch, M.; Rajagopolan, M. S.; Gantz, D. M.; Shuyarn, F.; Steimer, K.; Vyas, G. N.: In situ hybridization and immunocytochemistry for improved assessment of human immunodeficiency virus cultures. Am. J. Clin. Pathol. *88:* 673–680 (1987).
8 Farzadegan, H.; Polis, M. A.; Wolinsky, S. M.; Rinaldo, C. R., Jr.; Sninsky, J. J.; Kwok, S.; Griffith, R. L.; Kaslow, R. A.; Phair, J. P.; Polk, B. F.: Loss of human immunode-

ficiency virus type 1 (HIV-1) antibodies with evidence of viral infection in asymptomatic homosexual men. A report from the Multicenter AIDS Cohort Study. Ann. Intern. Med. *108:* 785 (1988).

9 Fauci, A. S.: The human immunodeficiency virus: Infectivity and mechanisms of pathogenesis. Science *239:* 617 (1988).

10 Goudsmit, J.; de Wolf, F.; Paul, D. A.; Epstein, L. G.; Lange, J. M.; Krone, W. J.; Speelman, H.; Wolters, E. C.; Van der Noordaa, J.; Oleske, J. M.: Expression of human immunodeficiency virus antigen (HIV-Ag) in serum and cerebrospinal fluid during acute and chronic infection. Lancet *ii:* 177 (1986).

11 Harper, M. E.; Marselle, I. M.; Gallo, R. C.; Wong-Staal, F.: Detection of lymphocytes expressing human T-lymphotropic virus type III in lymph nodes and peripheral blood from infected individuals by in situ hybridization. Proc. Natl. Acad. Sci. USA *83:* 772 (1986).

12 Henderson, A.; Ripley, S.; Heller, M.; Kieff, E.: Chromosome site for Epstein-Barr virus DNA in a Burkitt tumor cell line and in lymphocytes growth transformed in vitro. Proc. Natl. Acad. Sci. USA *80:* 1987–1991 (1983).

13 Kwok, S.; Mack, D. H.; Mullis, K. B.; Poiesz, B.; Ehrlich, G.; Blair, D.; Friedman-Kien, A.; Sninsky, J. J.: Identification of human immunodeficiency virus sequences by using in vitro enzymatic amplification and oligomer cleavage detection. J. Virol. *61:* 1690 (1987).

14 Langer, P. R.; Waldrop, A. A.; Ward, D. C.: Enzymatic synthesis of biotin-labeled polynucleotides: Novel nucleic acid affinity probes. Proc. Natl. Acad. Sci. USA *78:* 6633 (1981).

15 Langer-Safer, P. R.; Levine, M.; Ward, D. C.: Immunological method for mapping genes on Drosophilia polytene chromosomes. Proc. Natl. Acad. Sci. USA *79:* 4381 (1982).

16 Lawrence, J. B.; Singer, R. H.: Quantitative analysis of in situ hybridization methods for the detection of actin gene expression. Nucl. Acids Res. *13:* 1777 (1985).

17 Lawrence, J. B.; Singer, R. H.; Marselle, L. M.: Highly localized distribution of specific transcripts within interphase nuclei by in situ hybridization. Cell *57:* 493–502 (1989).

18 Lawrence, J. B.; Villnave, C. A.; Singer, R. H.: Sensitive, high resolution chromatin and chromosome mapping in situ: presence and orientation of two closely integrated copies of EBV in a lymphoid line. Cell *52:* 51–61 (1988).

19 Leary, J. J.; Brigati, D. J.; Ward, D. C.: Rapid and colorimetric method for visualizing biotin-labelled DNA probes hybridized to DNA or RNA immobilized on nitrocellulose: Bio-blots. Proc. Natl. Acad. Sci. USA *80:* 4045 (1983).

20 Mayer, K. H.; Stoddard, A. M.; McCusker, J.; Ayotte, D.; Ferriani, R.; Groopman, J. E.: Human T-lymphotropic virus type III in high-risk, antibody-negative homosexual men. Ann. Intern. Med. *104:* 194 (1986).

21 Miller, G.; Shape, T.; Coope, D.: Lymphoma in cotton-top marmosets after inoculation with Epstein-Barr virus: tumor incidence, histologic spectrum, antibody responses, demonstration of viral DNA, and characterization of viruses. J. Exp. Med. *145:* 948–967 (1977).

22 Monroe, J. E.; Andrews, C.; Sullivan, J. L.; Mulder, C.: Use of cytoplasmic dot-blot hybridization to detect human immunodeficiency virus RNA sequences in cultures of peripheral blood. J. Infect. Dis. *155:* 320 (1987).

23 Ou, C. Y.; Kwok, S.; Mitchell, S. W.; Mack, D. H.; Sninsky, J. J.; Krebs, J. W.; Feorino,

P.; Warfield, D.; Schochetman, G.: DNA amplification for direct detection of HIV-1 in DNA of peripheral blood mononuclear cells. Science *239:* 295 (1988).
24 Scott, G. B.; Buck, B. E.; Letterman, J. G.; Bloom, F. L.; Parks, W. P.: Acquired immunodeficiency syndrome in infants. N. Engl. J. Med. *310:* 76 (1984).
25 Shaw, G. M.; Hahn, B. H.; Arya, S. K.; Groopman, J. E.; Gallo, R. C.; Wong-Staal, F.: Molecular characterization of human T-cell leukemia (lymphotrophic) virus type III in the acquired immune deficiency syndrome. Science *226:* 1165–1171 (1984).
26 Shaw, G. M.; Harper, M. E.; Hahn, B. H.; Epsten, L. G.; Gajdusek, D. C.; Price, R. W.; Navia, B. A.; Petito, C. K.; O'Hara, C. J.; Groopman, J. E.: HTLV-III infection in brains of children and adults with AIDS encephalopathy. Science *227:* 177 (1985).
27 Singer, R. H.; Byron, K. S.; Lawrence, J. B.; Sullivan, J. L.: Detection of HIV-1 infected cells from patients using non-isotopic in situ hybridization: developing a rapid, sensitive diagnostic assay. Blood *14:* (in press, 1989).
28 Singer, R. H.; Lawrence, J. B.; Rashtchian, R. N.: Toward a rapid and sensitive in situ hybridization methodology using isotopic and non-isotopic probes; in Valentino, K., Eberwine, J., Barchas, J. (eds): In situ Hybridization: Application to Neurobiology, pp. 71–96 (Oxford University Press, New York 1987).
29 Singer, R. H.; Lawrence, J. B.; Villnave, C.: Optimization of in situ hybridization using isotopic and non-isotopic detection methods. Biotechniques *4:* 230 (1986).
30 Somasundaran, M.; Robinson, H. L.: Unexpectedly high levels of HIV-1 RNA and protein synthesis in a cytocidal infection. Science *242:* 1554–1557 (1988).
31 Tenner-Racz, K.; Racz, P.; Schmidt, H.; Dietrich, M.; Kern, P.; Louie, A.; Gartner, S.; Popovic, M.: Immunohistochemical, electron microscopic and in situ hybridization evidence for the involvement of lymphatics in the spread of HIV-1. AIDS *2:* 299–309 (1988).
32 Teo, C. G.; Griffin, B. E.: Epstein-Barr virus genomes in lymphoid cells: activation in mitosis and chromosomal location. Proc. Natl. Acad. Sci. USA *84:* 8473–8477 (1987).
33 Unger, E. R.; Budgeon, L. R.; Myerson, D.; Brigati, D. J.: Viral diagnosis by in situ hybridization. Description of a rapid simplified colorimetric method. Am. J. Surg. Pathol. *10:* 1 (1986).
34 Ward, J. W.; Holmberg, S. D.; Allen, J. R.; Cohn, D. L.; Critchley, S. E.; Kleinman, S. H.; Lenes, B. A.; Ravenholt, O.; Davis, J. R.; Quinn, M. G.; Jaffe, H.: Transmission of human immunodeficiency virus (HIV) by blood transfusions screened as negative for HIV antibody. N. Engl. J. Med. *318:* 473 (1988).

R. Singer, Department of Cell Biology, University of Massachusetts Medical School, 55 Lake Avenue North, Worcester, MA 01655 (USA)

Racz P, Haase AT, Gluckman JC (eds): Modern Pathology of AIDS and Other Retroviral Infections. Basel, Karger, 1990, pp 42–50

Detection of Lymphocytes Expressing Human Immunodeficiency Virus HIV-1 by in situ Hybridization with Oligonucleotide cDNA Probes[1]

Dominico Lazzaro[a], *Thomas Benter*[b], *Steven F. Josephs*[c,2]

[a]Virology Department, Columbia Medical Center, New York, N.Y., USA;
[b]Hannover University Medical School, Lehrte/Hannover, FRG;
[c]Laboratory of Tumor Cell Biology, National Cancer Institute, National Institutes of Health, Bethesda, Md., USA

Nucleic acid hybridization has found widespread application in genetic research, biomedical research and clinical diagnostics. In the last few years, in situ hybridization methodology in virological studies has become a powerful tool for the identification of the cell types implied in different infectious diseases [1–4]. Typically, for the detection of retrovirus RNA or DNA, radiolabeled transcripts are generated from proviral DNA cloned into riboprobe vectors.

For standard hybridization reactions in which radioisotope-labeled probes are annealed to a DNA or RNA sample [5], limitations exist due to radioisotope instability, disposal problems, and potential health hazards. The development of a nonradioactive DNA/RNA detection system to overcome these problems requires a method to label the nucleic acid probe and a sensitive and specific procedure to detect the annealed probe-target hybrid. The high binding constant and specificity of biotin for avidin or streptoavidin has been used in a variety of systems to localize and to detect a wide range of target molecules [1, 6, 7]. Langer et al. [6] and Brigati et al. [1] have described a series of biotin-labeled UTP and dUTP nucleotide analogues which contain biotin attached at the 5-position of the pyrimidine base. These modified

[1]This work was partially funded by the American Association for AIDS Research Grant No. 0005 to D.L.

[2]We thank Drs. R. C. Gallo and F. Wong-Staal for their support.

nucleotides are substrates for a variety of DNA and RNA polymerases and the biotin groups are accessible to avidin and/or antibody-conjugated enzymes after enzymatic incorporation into polynucleotide probes. Forster et al. [7] have synthesized a photoactivatable biotin reagent (photobiotin) which introduces biotin into nucleic acids upon irradiation with visible light. Recently, Pezzella et al. [4] employed cDNA probes modified by chemical insertion of an antigenic sulfone group in cytosine moieties which was visualized by a double antibody immunohistochemical reaction [4].

We introduce a method for enzymatic biotin labeling of cDNA oligonucleotide probes [6, 8] and comparative results are presented with radioactive labeled probes for the detection of lymphocytes expressing human immunodeficiency virus type 1 (HIV-1) by the use of in situ hybridization techniques.

Materials and Methods

Cells and Cell Lines

Cell lines H9 and infected H9/HTLV-IIIB [9] were resuspended at a density of 10^6/ml in RPMI medium containing 10% fetal calf serum. Cells were cytocentrifuged onto precleaned slides at a density of 1×10^5, air dried for 5 min, and then fixed for 1 min in 4% paraformaldehyde in phosphate-buffered saline (PBS) and then transferred to 70% ethanol. Preparations were stored at 4°C until used for hybridization.

Tissues

Fresh biopsy lymph nodes from AIDS-related complex (ARC) patients were minced and teased over a sterile stainless steel screen and the released cells cultivated in RPMI 1640 with 10% fetal calf serum for 24–48 h. The cells were cytocentrifuged, fixed, and stored as described above.

Oligonucleotide Probes

A positive strand oligonucleotide of 77 bases corresponding to the mRNA leader region of HIV-1 (+201 to +277 relative to the site of initiation of RNA synthesis), a 10 base primer complementary to the 3′ end of the 77 base oligonucleotide and a negative strand oligonucleotide of 51 base pairs (+5 to +55) were synthesized using a automated Applied Biosystems (Foster City, Calif.) DNA synthesizer. ^{35}S-labeled oligonucleotides were prepared by enzymatic synthesis. One picomole of the 77 base oligonucleotide and 5 pmol of the 10 base complementary primer were heated at 52°C for 10 min in 50 μl of 50 m*M* Tris-Cl pH 7.5, 5 m*M* $MgCl_2$, 5 m*M* 2-mercaptoethanol, with and then allowed to cool to room temperature. The solution of annealed primer template was made 30 μ*M* in ^{35}SdCTP and ^{35}SdATP and 100 μ*M* unlabeled dGTP and TTP. Five units of DNA polymerase I (Klenow fragment, Boehringer) were added and the reaction incubated at 22°C for 30 min. The reaction was terminated by boiling 2 min and the probe separated on a Nensorb 20 cartridge (Dupont).

Biotinylation of the 77 base probe was carried out enzymatically as described above except that biotin d-UTP (bio-11-UTP, Enzo Biochem) was used instead of the ^{35}S-labeled oligonucleotides and the reaction mixture was 30 μM dATP, dCTP, dGTP and bio-11-UTP, 50 mM Tris-Cl pH 7.5, 5 mM $MgCl_2$. The 51 base oligonucleotide was photobiotinylated with a photoprobe biotin kit (Vector) [7].

In situ Hybridization

Slide preparations were rinsed for 30 s in 2 × SSC (0.3 M NaCl, 0.03 M Na citrate) and acetylated in acetic anhydride/triethanolamine, pH 8.0 [3]. Slides were rinsed 30 s in 2 ×SSC and then dehydrated in ethanol. The hybridization mixture contained 50% formamide, 2 × SSC, 10 mM dithiothreitol, 1 mg/ml sheared salmon sperm DNA, 1 mg/ml *Escherichia coli* tRNA, 10% dextran sulfate, 2 mg/ml bovine serum albumin and alternatively the following concentrations of probes were added: 77 base oligonucleotide labeled with ^{35}S (10^8 dpm/ml; 50–100 ng) or with biotin (25–75 ng/ml), or 51 base photobiotinylated oligonucleotide (25–50 ng/ml).

The hybridization mixtures were heated at 90 °C for 10 min and quickly quenched on ice. The mixes were applied in 30-μl volumes, coverslips were mounted and sealed with rubber cement and denaturation accomplished by heating the slides for 5 min at 80 °C in a dry oven. The slides were immediately transferred onto a histology hot plate (Fisher) and hybridization was carried out for 3 h at 37 °C. Slides were rinsed thoroughly in 50% formamide, 2 × SSC, at 42 °C, followed by 6 rinses in 2 × SSC at room temperature and dehydrated by sequential rinsing for 2 min each in 50, 70, 95 and 100% ethanol.

Detection Systems

Cell preparations hybridized with ^{35}S-labeled probes were autoradiographed with NTB_2 nuclear track emulsion (Eastman) diluted 1:1 with distilled water. After exposure for 2–4 days at 4 °C, slides were developed and stained with Wright's and/or Giemsa stain.

Fig. 1. HIV-1-infected H9 cells were hybridized in situ with ^{35}S-labeled oligonucleotides as specific probes for HIV-1. Exposure time was 5 days. Giemsa stain ×90.

Fig. 2. Uninfected H9 cells were hybridized in situ with ^{35}S-labeled oligonucleotides as in figure 1. Wright's stain. ×110.

Fig. 5a,b. Lymphocytes derived from a node of an ARC patient hybridized in situ with ^{35}S-labeled oligonucleotides. Exposure time was 5 days. Wright's stain. ×220.

Fig. 6. Isolated cells from the same patient as cells of figure 5 after an in situ hybridization with photobiotinylated oligonucleotides specific for HIV-1 sequences. Detection method: indirect immunoperoxidase. Wright's stain. ×175. Note the distribution of the signal in the cytoplasm (see text).

Fig. 7. Cells isolated as in figure 5 and hybridized in situ with an enzymatically biotinylated specific for HIV-1 sequences. Detection method: indirect immunoperoxidase reaction. Wright's stain. ×175.

Fig. 8. Isolated cells from an ARC patient lymph node primary culture treated with DNase I (see text) and hybridized in situ with ^{35}S-labeled oligonucleotides specific for HIV-1 sequences. Exposure was for 5 days. Wright's stain. ×440.

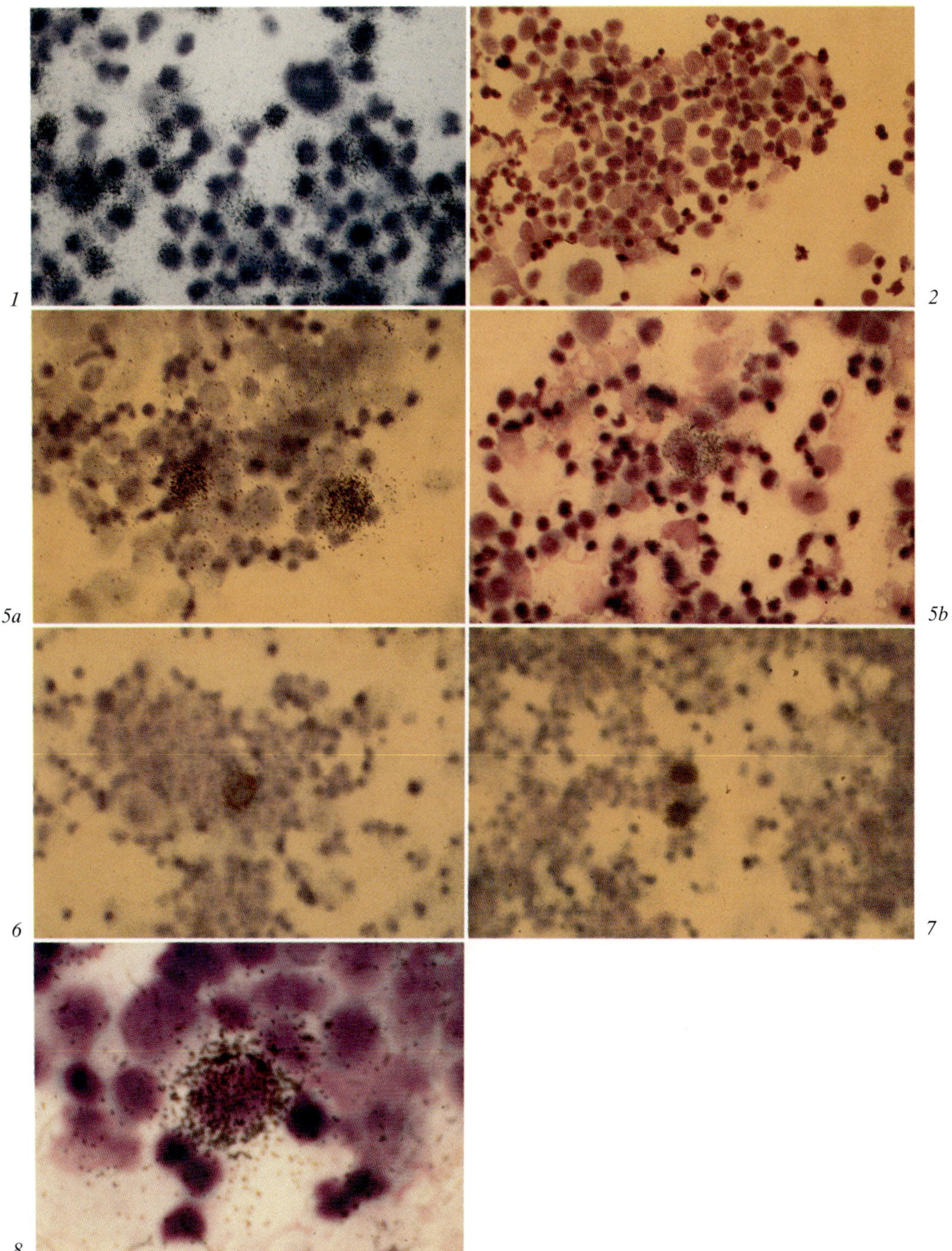

1
2
5a
5b
6
7
8

For cytopsin cell preparations hybridized with biotinylated oligonucleotides, the detection system used was an indirect immunoperoxidase procedure with rabbit antibiotin IgG and swine antirabbit peroxidase-labeled IgG. After the hybridization and washing procedures described above, the cells were allowed to react with 0.01% hydrogen peroxide in methanol for 30 min and washed in 1× PBS for 1 min then blocking buffer (1× PBS, 0.1% Triton X-100) 1 min, and 1×PBS 1 min. Slides were then treated with 100 μl of rabbit antibiotin IgG diluted 1:500 (Enzo Biochem, N.Y.) in 1× PBS at 37°C for 30 min, and washed in WB buffer (1 m*M* KOP_4, pH 6.5, 0.5 *M* NaCl, 0.5% Triton X-100, 1.0 m*M* EDTA, 0.1% BSA) for 5 min, then incubated with swine antirabbit peroxidase-labeled IgG diluted 1:50 in 1× PBS for 30 min at 37°C and washed in WB buffer with two changes of 5 min each. The slides were immersed in predetection buffer (100 m*M* Tris-Cl pH 8.8, 100 m*M* NaCl, 5 m*M* $MgCl_2$). The peroxidase was developed with a diaminobenzidine hydrochloride medium containing hydrogen peroxide according to the manufacturer's instructions (Enzo Biochem, N.Y.).

Controls consisted of cell spreads incubated with hybridization buffer only (without probe) under identical conditions, and by substituting the biotinylated oligonucleotide probes with biotinylatd pBR322 in the hybridization mix. The control uninfected lymphocytes tested were H9 cells and lymphocytes from normal donors.

Results

Specificity of in situ Hybridization for HIV-1 RNA

Specificity of the in situ hybridization method was assessed by the use of H9 cells, a neoplastic aneuploid cell line significantly resistant to the cytopathic effects of HIV-1. A highly significant labeling was observed over H9 cells productively infected with HIV-1 (H9/HTLV-IIIB). As shown in figure 1, after hybridization with ^{35}S-labeled oligonucleotide, essentially 100% of cells in this cell line were highly labeled after 3 days of autoradiographic exposure. By microscopy, 50–100 grains/cell were observed over the nucleus and cytoplasm. In contrast, cell preparations from uninfected H9 cells exhibited essentially no positive label when hybridized with radioactively labeled oligonucleotides as shown in figure 2. Very few grains were detected in uninfected H9 cell preparations treated with ^{35}S-oligonucleotide probes (fig. 2).

Similar specificity was obtained by the use of biotinylated oligonucleotides. No staining was detected in the control uninfected H9 cells (fig. 3) whereas a positive peroxidase staining was detectable in nearly 100% of infected H9 cells (fig. 4). The staining was confined mainly to the cytoplasm. When the infected H9 cells were hybridized with bacteriophage lambda-specific control riboprobe, with biotinylated pBR322 or with hybridization buffer alone no labeled cells were observed (data not shown).

Cell preparations of cultured lymph nodes from an ARC patient which were probed with ^{35}S-labeled and biotinylated oligonucleotides exhibited 5–

Fig. 3. Uninfected H9 cells after an in situ hybridization with photobiotinylated oligonucleotides as specific probes for HIV-1. Detection method: indirect immunoperoxidase reaction. Shown is a phase-contrast micrograph. ×130.

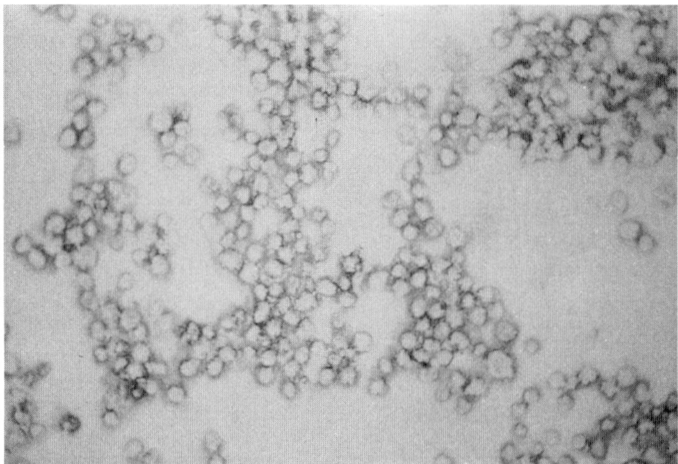

Fig. 4. Infected H9 cells after an in situ hybridization with photobiotinylated oligonucleotides and indirect immunoperoxidase reaction as in figure 3. Shown is a phase-contrast micrograph. ×130.

Table 1. HIV-1-specific oligonucleotides

A)	5′ CTC GCC TCT T 3′
B)	5′ CTG AAA GCG AAA GGG AAA CCA GAG CTC TCT CGA CGC AGG ACT CGG CTT GCT GAA GCG CCC GCA CGG CAA GAG GCG AG 3′
C)	5′ TCC CTA GTT AGC CAG AGA CGT CCC AGG CTC AGA TCT GGT CTA ACC AGA GAG 3′

A) Sequence of the 10 base oligonucleotide primer for template.
B) Used in enzymatic isotopic and nonisotopic labeling for HIV-1-specific probes.
C) The 51 base oligonucleotide used for photobiotinylation.

10% labeling of mononuclear cells. The same percentage was pointed out by a previous immunocytochemical screening of the same samples [A. Minassian, pers. commun.]. After a 4-day exposure following hybridization with the ^{35}S-labeled oligonucleotide probe, 100–200 grains were localized over infected cells (fig. 5a, b). The grains were located over the nucleus and the cytoplasm and in the immediate periphery, possibly because of leakage of cell material [3].

When biotinylated probes were used, as with the HIV-1-infected H9 cells, the positive mononuclear cells showed a strong peroxidase staining mainly located in the cytoplam. Very few cells showed nuclear staining (fig. 6, 7). The HIV-1-positive cells were similar in size and morphological detail to lymphoid cells (fig. 5a, 6, 7); however, some cells were seen which had a smaller nucleus/cytoplasm ratio (fig. 5b) than expected of resting lymphocytes. Cell preparations of the same patient hybridized with ^{35}S-labeled phage lambda-specific control riboprobe (Promega Biotec.) or with biotinylated pBR322 nick-translated probe were consistently negative (data not shown). Similarly, mononuclear cell preparations from normal individuals were negative when hybridized with ^{35}S-labeled or biotinylated the oligonucleotides specific for HIV-1 RNA sequences (data not shown).

Specificity for Viral RNA

Specificity for viral RNA as opposed to DNA was demonstrated by hybridization of radioactively labeled and biotinylated oligonucleotides to H9-infected cells after digestion with either RNase (RNase A at 100 μg/ml; RNase T_1 at 1 μg/ml) or DNase I (20 μg/ml) at 37°C for 18 h. H9/HTLV-IIIB cells treated with DNase prior to hybridization exhibited significant label at a similar level to that described for the untreated positive samples (data not shown). The ARC patient's lymph node cells treated with DNase I gave similar results, comparable to that seen in untreated samples (fig. 8). However, the infected cells which were treated with RNase before hybridization exhibited essentially no labeling (data not shown). These observations indicate that under the hybridization conditions used, the label was specific for viral RNA and not integrated or unintegrated HIV-1 DNA.

Discussion

In situ hybridization was used to detect cells expressing HIV-1 RNA in an HIV-1-infected cell line and from cultured lymph node samples from ARC patients by the use of isotopic and nonisotopic detection methods.

This study describes a technique which allows the use of cDNA oligonucleotides as suitable probes for the detection of HIV-1-infected cells by in situ hybridization of isotopic and nonisotopic probes. We presented comparative results obtained by the use of enzymatically ^{35}S-labeled or biotinylated probes and photobiotinylated probes. Infected cells were clearly visualized by the use of isotopic and nonisotopic labels.

The hybridization was specific for viral RNA sequences and the same percentage of infected cells was detected in different samples by the use of either radioactive or biotinylated oligonucleotide probes. A comparison of isotopic versus nonisotopic labeled oligonucleotides revealed the only remarkable difference concerned nuclear positivity. Our results confirmed the observations of Harper et al. [3], that positively labeled mononuclear cells presented grains located over the nucleus and the cytoplasm. On the other hand, with nonisotopic detection systems, this nuclear positivity was observed only in few cells similar to the findings of Pezzella et al. [4]. The background and specificity problems inherent to the use of biotinylated probes [10] for in situ hybridization seemed to be minimized by the addition of Triton X-100 to prehybridization washed and buffer.

In recent studies, several authors reported a marked nonspecific labeling over section and cultured cell preparations using ^{35}S-labeled riboprobes or synthetic oligonucleotides [11, 12]. Since this nonspecific labeling was independent of the size and sequence of probes and it could not be reduced by more stringent washing conditions nor by intense RNase digestion, the authors concluded that the substitution of an O by an S in the phosphate groups of the nucleotides led to a change in the chemical reactivity which probably resulted in covalent binding to non-RNA macromolecules.

According to our results and considering the length of our synthetic probes, we can exclude the formation of any probe network [10] that could have significantly contributed to the signal observed over the nucleus of infected cells hybridized with ^{35}S-labeled oligonucleotides. Furthermore, although previous studies have shown the copy number of integrated and unintegrated HIV-1 DNA in the chronically infected cells is variable [13], the cells treated with DNase prior to hybridization showed a similar nuclear label [3]. Thus, the nuclear positivity observed with ^{35}S-labeled probes, riboprobes or oligonucleotides does not seem to be due to specific hybridization to nuclear viral DNA.

Conclusion

In summary, our results support previous studies [1–4] and confirm a preliminary immunocytochemical analysis of the same samples [A. Minassian, pers. commun.] thereby proving the efficacy of synthetic oligonucleotides as probes for in situ hybridization [8, 14].

The techniques described should be advantageous for both basic and clinical investigations [15, 16]. The procedures described here show a sensitivity of detection using biotinylated and photobiotinylated oligonucleotides that favorably compares to the levels achieved with radiolabeled probes [1–3]. Rapid preparation of radiolabeled oligonucleotide probes was possible which resulted in very clean background with hybridization equal to or better than that resulting from use of longer riboprobes. Because of the small size of the probes, it will be of interest to test these techniques on processed tissue samples including those treated with cross-linking fixatives [2]. It will also be of interest to perform in situ hybridization studies at the ultrastructural level by the use of biotinylated oligonucleotides and colloidal gold-conjugated antibiotin antibodies and/or avidin as detection methods [17].

References

1 Brigati, D.J.; Myerson, D.; Leary, J.J.; Spalholz, B.: Travis, S.Z.; Fong, C.K.Y.; Hsiung, G.D.; Ward, D.C.: Detection of viral genomes in cultured cells and paraffin-embedded tissue sections using biotin-labeled hybridization probes. Virology *126:* 32–50 (1983).

2 Haase, A.; Brahic, M.; Stowring, L.; Blum, H.: Detection of viral nucleic acids by in situ hybridization; in Haase, Brahic, Methods in virology, vol. II, pp. 189–225 (Academic Press, New York 1984).

3 Harper, M. E.; Marselle, L. M.; Gallo, R. C.; Wong-Staal, F.: Detection of lymphocytes expressing human T-lymphotropic virus type III in lymph nodes and peripheral blood from infected individuals by in situ hybridization. Proc. natn. Acad. Sci. USA *83:* 772–776 (1986).
4 Pezzella, M.; Pezzella, F.; Galli, C.; Macchi, B.; Verani, P.; Sorice, F.; Baroni, C. D.: In situ hybridization of human immunodeficiency virus (HTLV III) in cryostat sections of lymph nodes of lymphadenopathy syndrome patients. J. med. Virol. *22:* 135–142 (1987).
5 Maniatis, T. E. F.; Fritsch, E. F.; Sambrook, J.: Molecular cloning, a laboratory manual (Cold Spring Harbor Laboratory Publications, Cold Spring Harbor 1982).
6 Langer, P. R.; Waldrop, A. A.; Ward, D. C.: Enzymatic synthesis of biotin-labeled polynucleotides: novel nucleic acid affinity probes. Proc. natn. Acad. Sci. USA *78:* 6633–6637 (1981).
7 Forster, A. C.; McInnes, J. L.; Skingle, D. C.; Symmons, R. H.: Nonradioactive hybridization probes prepared by the chemical labelling of DNA and RNA with a novel reagent, photobiotin. Nucl. Acids Res. *13:* 745–761 (1985).
8 Uhl, G. R.; Zingg, H. H.; Haebner, J. F.: Vasopressin mRNA in situ hybridization: localization and regulation studied with oligonucleotide cDNA probes in normal and Brattleboro rat hypothalamus. Proc. natn. Acad. Sci. USA *82:* 5555–5559 (1985).
9 Popovic, M.; Sarngadharan, M. G.; Read, E.; Gallo, R. C.: Detection, isolation, and continuous production of cytopathic retroviruses (HTLV III) from patients with AIDS and pre-AIDS. Science *224:* 497–500 (1984).
10 Lawrence, B. J.; Singer, R. H.: Quantitative analysis of in situ hybridization methods for the detection of actin gene expression. Nucl Acids Res. *13:* 1777–1799 (1985).
11 Brandtlow, C. E.; Heumann, R.; Schwab, M. E.; Thoenen, H.: Cellular localization of nerve growth factor synthesis by in situ hybridization. Eur. molec. Biol. Org. J. *6:* 891–899 (1987).
12 Berger, C. N.: In situ hybridization of immunoglobulin-specific RNA in single cells of the B lymphocyte lineage with radiolabelled DNA probes (erratum: Eur. molec. Biol. Org. J. *5:* 2747). Eur. molec. Biol. Org. J. *5:* 83–93 (1986).
13 Shaw, G. M.; Hahn, B. H.; Arya, S. K.; Groopman, J. E.; Gallo, R. C.; Wong-Staal, F.: Molecular characterization of human T-cell leukemia (lymphotropic) virus type III in the acquired immune deficiency syndrome. Science *226:* 1165–1171 (1984).
14 Largent, B. L.; Jones, D. T.; Reed, R. R.; Pearson, R. C. A.; Snyder, S. H.: G protein mRNA mapped in rat brain by in situ hybridization. Proc. natn. Acad. Sci. USA *85:* 2864–2868 (1988).
15 Martell, M.; Le Gall, I.; Millasseau, P.; Dausset, J.; Cohen, D.: Use of synthetic oligonucleotides for genomic DNA dot hybridization to split the DQw3 haplotype. Proc. natn. Acad. Sci. USA *85:* 2682–2685 (1988).
16 Gendelmann, H. E.; Moench, T. R.; Narayan, O.; Griffin, D. E.; Clements, J. E.: A double labeling technique for performing immunocytochemistry and in situ hybridization in virus-infected cell cultures and tissues. J. virol. Meth. *11:* 93–103 (1985).
17 Oakes, M. I.; Clark, M. W.; Henderson, E.; Lake, J. A.: DNA hybridization electron microscopy: ribosomal RNA nucleotides 1392–1407 are exposed in the cleft of the small subunit. Proc. natn. Acad. Sci. USA *83:* 275–279 (1986).

Dominico Lazzaro, MD, Virology Department, Columbia Medical Center,
650 West 168th Street, Room 1516 West Black Building, New York, NY 10032 (USA)

Racz P, Haase AT, Gluckman JC (eds): Modern Pathology of AIDS and Other Retroviral Infections. Basel, Karger, 1990, pp 51–61

Infection of Bone Marrow-Derived Dendritic Cells with HIV

Steven Patterson, Steven E. Macatonia, Jacqueline Gross, Stella C. Knight

MRC Clinical Research Centre, Harrow, Middx., UK

HIV-infected patients with generalised lymphadenopathy or with AIDS show a depletion of T lymphocytes, bearing the CD4 marker, and this is believed to be instrumental in the development of disease. However, other cells in the immune system are also susceptible to infection and this may contribute to the development of immunosuppression. For example, HIV replicates in activated macrophages [1]. Infection and destruction of the follicular dendritic cells (DC) of the B-dependent areas of lymph nodes [2] occurs and this may have functional implications for the development of B cell memory [3].

Langerhans' cells may be grossly infected with HIV in AIDS patients [4] and this probably accounts for the observed reduction in the number of MHC class II positive cells in the skin [5]. However, it is not yet clear whether this latter finding reflects destruction of Langerhans' cells or the down-regulation of their MHC class II antigens. Langerhans' cells form part of a population of bone marrow derived DC which are found in small numbers in most tissues of the body (<0.5%) and constitute a network of cells able to acquire and present antigens. Particularly after exposure to antigen they travel as veiled cells in the afferent lymphatics to the lymph nodes and become the interdigitating cells or DC of the T-dependent areas [6] (i.e. a population distinct from the follicular DC of lymph nodes which are derived in situ from fibroblastic elements [7]. DC of the T-dependent areas are able to form clusters with and activate T cells in both primary and secondary immune responses [8, 9].

Here we describe the sensitivity of normal DC isolated from human peripheral blood to infection with HIV and the immunosuppressive effects of this infection in vitro on lymphocyte stimulation. We also present evidence that DC are infected in vivo during HIV infection and discuss the functional relevance of these findings. The major conclusion is that a compromised antigen-presenting system may be a fundamental defect in HIV infection which is instrumental in producing immunosuppressive effects in AIDS.

Infection of Dendritic Cells in vitro

DC can be enriched from human peripheral blood using a technique in which the mononuclear cells are incubated overnight on plastic to remove a major proportion of the macrophages, and the non-adherent cells are then centrifuged over a metrizamide gradient (14.25% w/v). The combination of cell shrinkage and density using this hypertonic gradient facilitates the isolation of a preparation in which 30–50% of the cells are DC and most of the remaining cells are macrophages [11, 12]. When these cells were exposed to the IIIB strain of HIV and examined after only 3–5 days in culture by electron microscopy, many DC, but no macrophages, were grossly infected with virus (fig. 1) [13, 14]. The budding of the virus from the cell surface is in contrast with the form of growth reported in macrophages activated by colony-stimulating factor I where the virus was observed budding into the Golgi region [1]. The use of electron microscopy provides a method for secure identification of DC from their large size and characteristic morphology with their pale cytoplasm which is relatively devoid of organelles except for polyribosomes, scarce cytoplasmic vacuoles and the presence of veiled projections. Even in the rare DC showing evidence of a phagocytic history, the characteristic pale cytoplasm distinguishes them from the macrophages. Identification of human DC by other means can be problematical because of the lack of a single exclusive marker. However, these DC lack many monocyte markers (eg. CD14) and express high levels of all the class II MHC molecules, including HLA-DQ which is poorly expressed on macrophages [12, 16].

A major receptor for the virus is believed to be CD4 [15]. By immunogold labelling, DC were found to express low levels of membrane CD4 which was up-regulated by γ-interferon in vitro (fig. 2, 3) [Patterson et al., in preparation]. This suggests that when immune stimulation occurs and there is a resultant release of γ-interferon, one consequence could be the expression of increased receptor for virus and consequentially a greater susceptibility of DC to infection. The expression of CD4 and its enhancement with γ-interferon has also been reported in Langerhans' cells [17].

Evidence of virus infection can be obtained by immunolabelling or in situ hybridization techniques but the unequivocal identification of infected DC by

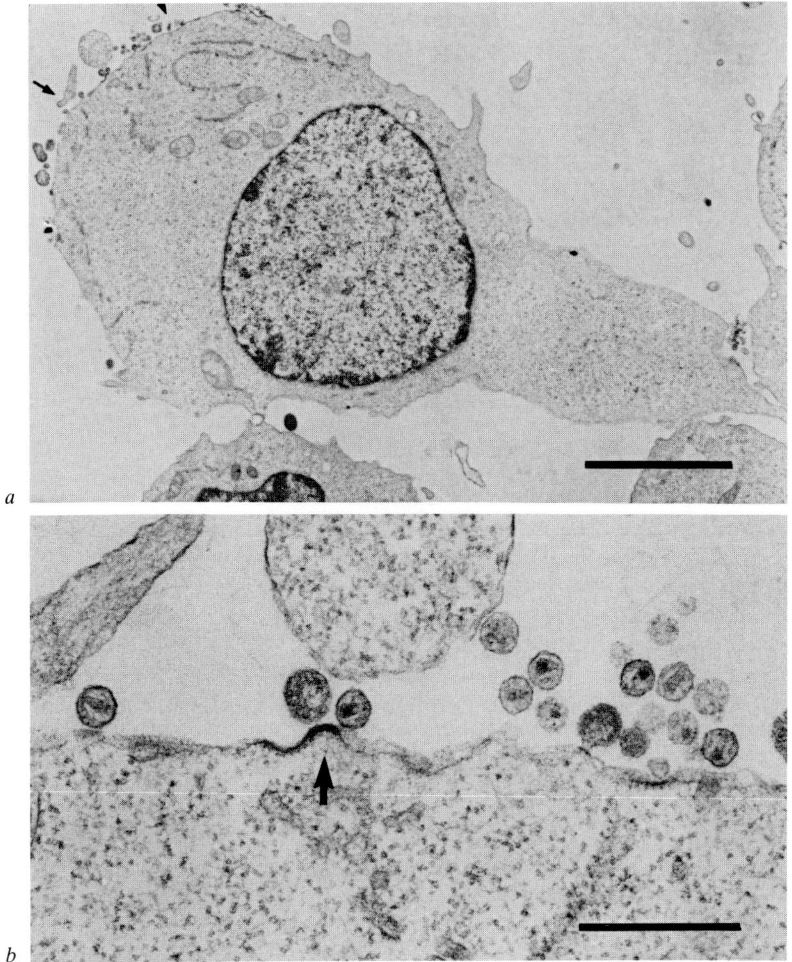

Fig. 1. *a* Electron micrograph of a DC 5 days after infection in vitro with HIV. Bar = 5 μm. *b* Higher magnification of the area indicated by the arrows in *a*. Virus particles are on the cell surface and particle in the process of budding through the plasma membrane may be observed (arrow). Bar = 500 nm.

light microscopy is difficult (fig. 4). In order to overcome this problem, cell suspensions in 0.01% sodium azide were labelled for 30 min on ice with a mixture of four monoclonal antibodies recognising CD19, CD14, CD16 and CD5 determinants which are specific for B cells, monocytes, NK cells and T cells respectively. After washing the cells they were adsorbed onto 0.01%

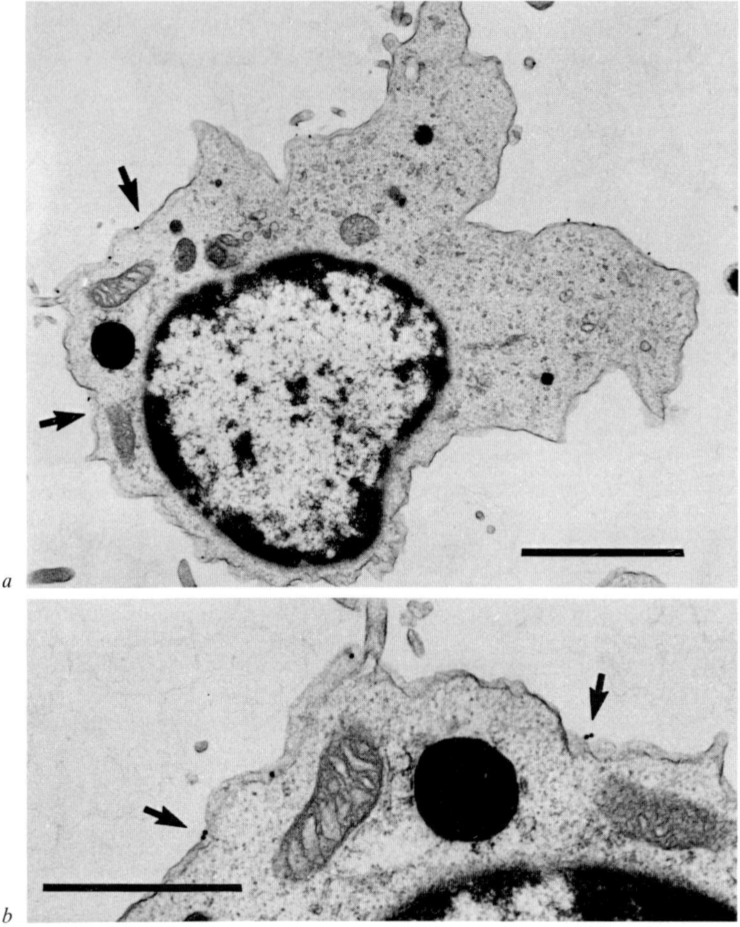

Fig. 2. *a* Immunogold labelling for CD4 on a DC. Bar = 2 μm. *b* A higher magnification of the area indicated by the arrows in *a*. Arrows indicate binding of colloidal gold particles. Bar = 1 μm.

poly-*L*-lysine coated slides and fixed with 4% paraformaldehyde in 0.1 *M* phosphate buffer pH 7.2 for 30 min. Binding of antibody was detected by the APAAP technique [18]. DC were characterised by their morphology plus absence of staining. The specificity of this technique was confirmed by adding purified DC to lymphocyte preparations and showing an increase in these large negative cells as a consequence [19]. The presence of HIV nucleic acid

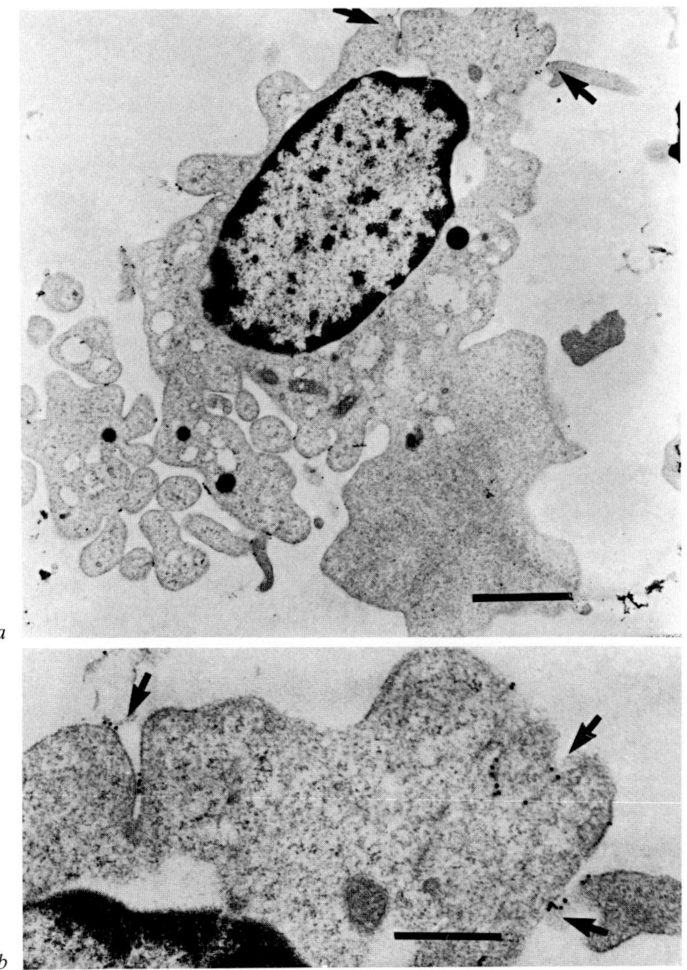

Fig. 3. *a* Immunogold labelling for CD4 on a DC after overnight incubation with γ-interferon (150 units/ml). Bar = 2 μm. *b* Higher magnification of the area indicated in *a*. Arrows indicate binding of colloidal gold particles. Bar = 1 μm.

was then detected in these same preparations by in situ hybridization [20] by a technique already described [21, 22] except that the probe was denatured after applying the hybridization mixture to the slides by heating to 95 °C for 6 min. This modification was introduced to detect viral DNA in addition to RNA and was necessary in order to increase the virus-specific signal as there was a marked loss of RNA labelling after immunolabelling which presumably re-

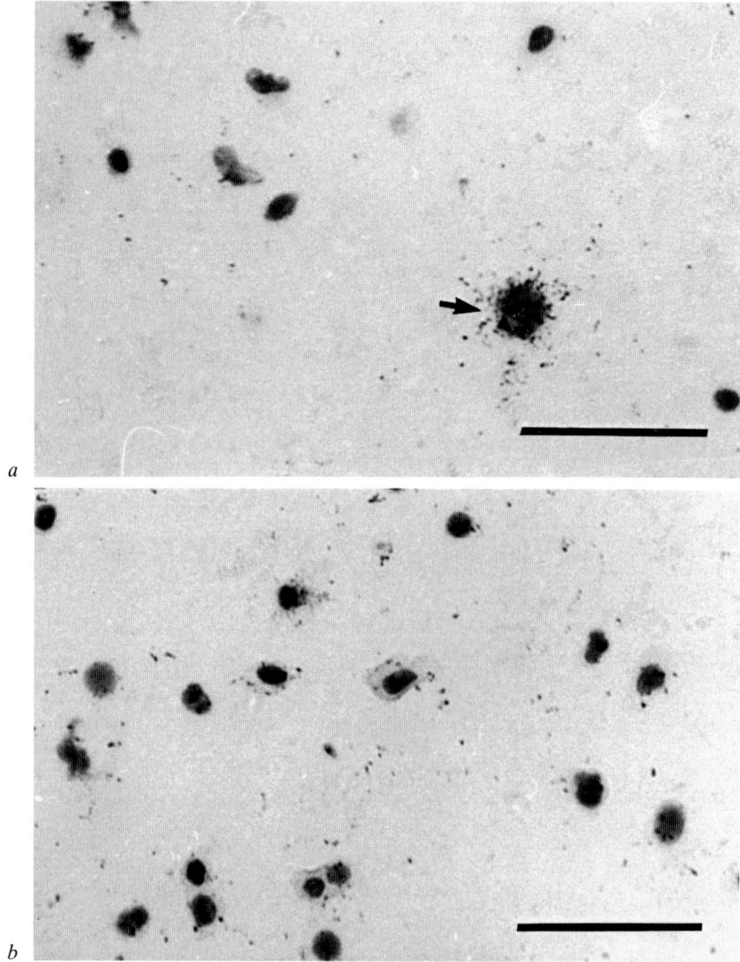

Fig. 4. a In situ hybridization of a DC preparation 10 days after in vitro infection with HIV. On morphological criteria alone it is difficult to identify unequivocally DC. Arrows indicate an infected cell. Bar = 20 μm. *b* A control uninfected DC preparation processed for in situ hybridization. Bar = 20 μm.

flects RNase activity in the antisera used. DC containing HIV genome could then be unequivocally identified by this double labelling procedure (fig. 5).

The functional consequences of in vitro infection of DC on the development of in vitro responses in syngeneic lymphocytes have been described [19, 23]. DC were infected with HIV and 2 days later pulsed with either influenza virus or concanavalin A and then incubated with lymphocytes in the hanging drop culture system. Three days later a significant reduction in the lymphocyte proliferative response was observed in cultures containing HIV-infected DC. Examination of infected antigen-stimulated cultures by in situ hybridization revealed that about 2% of the lymphocyte population had become infected with HIV. The intimate association of DC with lymphocytes during antigen presentation may facilitate the secondary infection of lymphocytes. Thus it was not clear whether the suppression of the in vitro lymphocyte proliferative response was due to the infection of the antigen-presenting DC or lymphocytes or both. To resolve this question, 2′3′-dideoxyadenosine was added to the infected DC preparation immediately prior to addition of antigen and co-culturing with lymphocytes. This reagent is a potent inhibitor of virus reverse transcriptase and, as shown by in situ hybridization, prevented secondary infection of lymphocytes. Although part of the inhibiting effect of the HIV was lost when T cell infection was prevented, the major inhibitory effects were still seen. Suppression of lymphocyte proliferation by both the direct infection of DC and by the DC acting as a reservoir of virus which was handed on to some T cells was therefore occurring.

In contrast to the blocking effect on lymphocyte stimulation that is caused by infection of DC with HIV, we have shown that the HIV-infected DC are able to initiate efficient responses to the HIV antigen itself in syngeneic lymphocytes in culture. These experiments used a system of lymphocyte stimulation carried out in 20 μl hanging drop cultures [26] which allow efficient presentation of viral antigens to lymphocytes by DC to produce primary proliferative and cytotoxic responses [10]. In initial experiments, mouse DC exposed to influenza virus in vitro showed no evidence of infection of DC but stimulated primary proliferative and cytotoxic syngeneic T lymphocyte responses. After 5 days in culture the stimulated lymphocytes caused specific lysis of virus-infected target cells. We have now shown that a similar system using human DC enriched from normal peripheral blood and infected with HIV in vitro caused primary proliferation and cytotoxic responses which killed virus-infected syngeneic T cell blasts [Macatonia et al., in preparation]. Production of primary responses to HIV were also described in cells stimulated with HIV-infected 'PHA blasts' [27]. Such cultures could also result in primary responses to HIV due to the infection of DC with the virus resulting in presentation of antigens to the lymphocyte responder population by a mechanism similar to that we have described [10].

In conclusion, in vitro studies with normal DC and HIV show that these cells are extremely susceptible to infection with virus. As many as 50% of the

DC present in mitogen-stimulated cultures were infected after 3 days and budding of virus particles from the surface of these cells was observed. This infection blocks the capacity of these cells to present mitogen or other antigen to lymphocytes. In contrast, HIV-infected DC can efficiently present antigens of the infecting virus to syngeneic T cells.

Infection of Dendritic Cells in vivo

The infection of Langerhans' cells of the skin in AIDS patients has already been described [4]. In addition, in AIDS patients there was a loss of cells with high levels of class II MHC antigens in the low density population of peripheral blood mononuclear cells isolated using hypertonic metrizamide [24]. One report also suggested that DC from peripheral blood of an AIDS patient were infected with HIV [25]. We have now confirmed the presence of virus in the DC of HIV seropositive individuals using the combined immunolabelling and in situ hybridization technique. DC, identified by the absence of the red alkaline phosphatase reaction product, were found to contain virus genome as shown by hybridization to a ^{35}S-labelled HIV DNA probe and subsequent autoradiography (fig. 6).

A block in the capacity of the low density DC to stimulate allogeneic mixed leukocyte reactions has been reported in AIDS [24]. The T cells of these patients also showed low responses to stimulation with normal allogeneic DC. These findings indicated that in AIDS there may be a defect in both stimulator and responder function involving DC and T lymphocytes respectively. However, we have observed changes in DC early in HIV infection which precede changes in the T lymphocytes. Reductions in DC numbers in

---

Fig. 5. Combined in situ hybridization and immunolabelling by the APAAP technique of a DC preparation 10 days after in vitro infection with HIV. Non-DC are stained red whilst DC are pale pink. Filled arrows indicate two infected DC. Open arrow indicates an infected non-DC. Bar = 20 μm.

Fig. 6. Combined in situ hybridization and immunolabelling by the APAAP technique of a DC preparation from an HIV seropositive individual. Non-DC are stained red whilst DC are pale pink. Filled arrow indicates an infected DC. Open arrows indicate two uninfected DC. Bar = 20 μm.

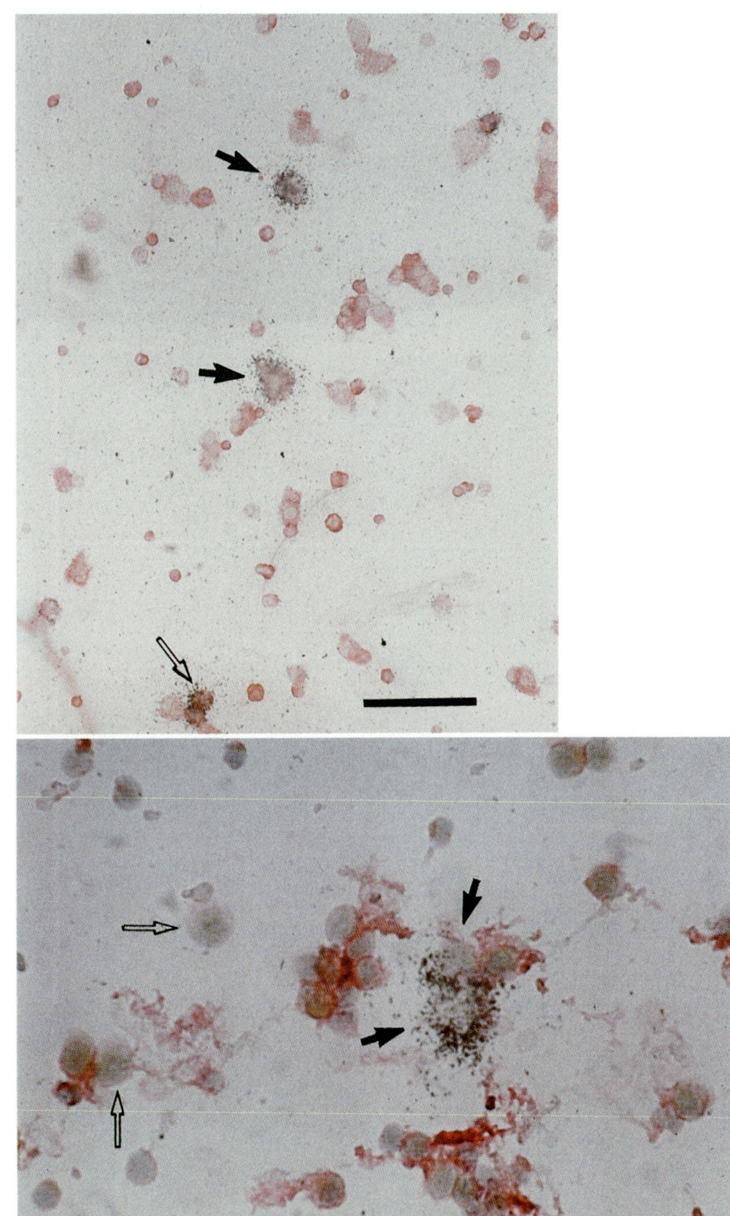

5

6

peripheral blood were seen in HIV-positive individuals who were asymptomatic and in whom the T cell numbers were still normal [Macatonia et al., in preparation]. In addition, the DC that remained were deficient in their capacity to stimulate proliferation of normal allogeneic lymphocytes although their T cells still responded well to stimulation with normal allogeneic DC. The functional deficit on DC was again present before any change in T cell function could be detected.

In conclusion, using peripheral blood from HIV-infected individuals, our double labelling studies confirm that DC are infected with HIV. The loss of circulating DC may be an early event on HIV infection. This could be related to the development of patient immune responses to virus-infected cells in vivo. Infected DC were shown to elicit potent primary cytotoxic lymphocyte responses in syngeneic lymphocytes in vitro. In HIV-infected individuals, many DC were lost but the DC that remained were compromised since they failed to stimulate allogeneic lymphocyte responses. The infection of DC and their loss in vivo, which preceded changes in T cell defects, might be an underlying cause of immunosuppressive effects in HIV infection. Loss of adequate accessory cell function might result in the inability of the antigen-presenting system to cause activation of immune responses since there is evidence that presentation of antigen by DC may be a prerequisite for the activation of resting T lymphocytes [10, 28].

References

1 Gendelman, H. E.; Orenstein, J. M.; Martin, M. A.; Ferrua, C.; Mitra, R.; Phipps, T.; Wahl, L. A.; Lane, H. C.; Fuci, A. S.; Burke, D. S.; Skillman, D.; Meltzer, M. S.: Efficient isolation and propagation of human immunodeficiency virus on recombinant idoxy-stimulating factor 1 treated monocytes. J. Exp. Med. *167:* 1428–1441 (1988).

2 Tenner-Racz, K.; Racz, P.; Gartner, S.; Ramsauer, J.; Dietrich, M.; Gluckman, J. C.; Popovic, M.: Ultrastructural analysis of germinal centers in lymph nodes of patients with HIV-1-induced persistent generalized lymphadenopathy: evidence for persistence of infection. Prog. AIDS Pathol. *1:* 29–40 (1989).

3 Klaus, G. G. B.; Humphrey, J. H.; Kunkl, A.; Dongworth, D. W.: The follicular dendritic cells: its role in antigen-presentation in the generation of immunological memory. Immunol. Rev. *53:* 3–28 (1980).

4 Tschachler, E.; Groh, U.; Popovic, M.; Mann, D.; Konrad, K.; Safai, B.; Eron, L.; Veronesi, F.; Wolff, K.; Stingl, G.: Epidermal Langerhans' cells – a target for HTLV-III/LAV infection. J. Invest. Dermatol. *88:* 233–237 (1987).

5 Belsito, D. V.; Senchez, M. R.; Baer, R. L.; Valentine, F.; Thorbecke, G. J.: Reduced Langerhans' cell 1a antigen and ATPase activity in patients with acquired immunodeficiency syndrome. N. Engl. J. Med. *310:* 1279–1282 (1984).

6 Silberberg-Sinakin, I.; Gigli, I.; Baer, R. L.; Thorbecke, G. J.: Langerhans' cells: role in contact hypersensitivity and relationship to lymphoid dendritic cells and macrophages. Immunol. Rev. *53:* 203–237 (1980).

7 Villena, A.; Zapata, A.; Rivera-Pomar, J. M.; Barrutia, M. G.; Fonfria, I. J.: Structure of the non-lymphoid cells during the postnatal development of the rat lymph nodes. Cell Tissue Res. *229:* 219–232 (1983).

8 Steinman, R. M.; Inaba, K.; Schuler, G.; Witmer, M.: Stimulation of the immune response: contributions of dendritic cells; in Steinman, R. M.; North, R. J. (eds): Mechanisms of Host Resistance to Infectious Agents, Tumours and Allografts, pp. 71–97 (Rockefeller Press, New York 1986).

9 Austyn, J. M.: Lymphoid dendritic cells. Immunology *62:* 161–170 (1987).

10 Macatonia, S. E.; Taylor, P.; Knight, S. C.; Askonas, B.: Primary stimulation by dendritic cells induces anti-viral proliferative and cytotoxic T cell responses in vitro. J. Exp. Med. *169:* 1255–1264 (1989).

11 Knight, S. C.; Farrant, J.; Bryant, A.; Edwards, A. J.; Burman, S.; Lever, A.; Clarke, J.; Webster, A. D. B.: Non-adherent, low-density cells from human peripheral blood contain dendritic cells and monocytes, both with veiled morphology. Immunology *59:* 595–603 (1986).

12 Knight, S. C.; Fryer, P.; Griffiths, S.; Harding, B.: Class II histocompatibility antigens in human dendritic cells. Immunology *61:* 21–27 (1987).

13 Patterson, S.; Knight, S. C.: Susceptibility of human peripheral blood dendritic cells to infection with human immunodeficiency virus. J. Gen. Virol. *68:* 1177–1181 (1987).

14 Knight, S.; Patterson, S.: Effect of human immunodeficiency virus on dendritic cells isolated from human peripheral blood; in Dupont, B. (ed.): Proc. 10th Int. Histocompatibility Workshop (Springer, Berlin, in press).

15 Dalgleish, A. G.; Beverley, P. C. L.; Clapham, P. R.; Crawford, D. H.; Greaves, M. F.; Weiss, R. A.: The CD4 (T4) antigen in an essential component of the receptor for the AIDS retrovirus. Nature (Lond.) *312:* 763–767 (1984).

16 Brooks, C. F.; Moore, M.: Differential MHC class II expression on human peripheral blood monocytes and dendritic cells. Immunology *63:* 303–311 (1988).

17 Walsh, L. J.; Parry, A.; Scholes, A.; Seymour, G. T.: Modulation of CD4 antigen expression on human gingival Langerhans' cells by gamma-interferon. Clin. Exp. Immunol. *70:* 379–385 (1987).

18 Mason, D. Y.: Techniques of Immunocytochemistry, pp. 25–40 (Academic Press, New York 1985).

19 Macatonia, S. E.; Patterson, S.; Knight, S. C.: Suppression of immune responses by dendritic cells infected with HIV. Immunology *67:* 285–289 (1989).

20 Hahan, B. H.; Shaw, G. M.; Arya, S. K.; Popovic, M.; Gallo, R. C.; Wong-Staal, F.: Molecular cloning and characterisation of the HTLV-III virus associated with AIDS. Nature (Lond.) *312:* 166–169 (1984).

21 Rigby, P. W. J.; Dieckmann, M.; Rhodes, G.; Berg, P.: Labelling deoxyribonucleic acid to high specific activity in vitro by nick translation with DNA polymerase. J. Mol. Biol. *113:* 237–251 (1977).

22 Patterson, S.; Gross, J.; Webster, A. D. B.: DNA probes bind non-specifically to eosinophils during in situ hybridization: carbol chromotope blocks binding to eosinophils but does not inhibit hybridization to specific nucleotide sequences. J. Virol. Methods *23:* 105–109 (1989).

23 Knight, S. C.; Macatonia, S. E.: Dendritic cells and viruses. Immunol. Lett. *19:* 177–182 (1988).
24 Eales, L. J.; Farrant, J.; Herbert, M.; Pinching, A. J.: Peripheral blood dendritic cells in persons with AIDS and AIDS-related complex: loss of high intensity class II antigen expression and function. Clin. Exp. Immunol. *71:* 423–427 (1988).
25 Ranki, A.; Krohn, M.; Allain, J.; Franchini, G.; Valle, S. L.; Antonen, J.; Leuther, M.; Krohn, K.: Long latency precedes overt seroconversion in sexually transmitted human immunodeficiency virus infection. Lancet *ii:* 589–593 (1987).
26 Knight, S. C.: Lymphocyte proliferation assays; in Klaus, G. G. B. (ed.): Lymphocytes: A Practical Approach, pp. 189–207 (IRL Press, Oxford 1988).
27 Hoffenbach, A.; Langlade-Demoyen, P.; Dadaglio, G.; Vilmer, E.; Michel, F.; Mayaud, C.; Autran, B.; Plata, F.: Unusually high frequency of HIV-specific cytotoxic T lymphocytes in humans. J. Immunol. *142:* 452–462 (1989).
28 Inaba, K.; Steinman, R. M.: Resting and sensitised T lymphocytes exhibit distinct stimulatory (antigen-presenting cell) requirements for growth and lymphokine release. J. Exp. Med. *160:* 1717–1735 (1984).

Dr. Stella C. Knight, Clinical Research Centre, Watford Road,
GB–Harrow, Middx. HA13UJ (UK)

Racz P, Haase AT, Gluckman JC (eds): Modern Pathology of AIDS and Other Retroviral Infections. Basel, Karger, 1990, pp 62–68

Fine Analysis of HIV-1-RNA Detection by in situ Hybridization[1]

J. G. Fournier[a], *S. Prevot*[b], *J. Audouin*[b], *A. Le Tourneau*[b], *P. Lebon*[2], *J. Diebold*[b,2]

[a]INSERM U. 43, Hôpital St. Vincent de Paul, et
[b]Service Central 'J. Delarue' d'Anatomie et de Cytologie Pathologiques, Hôtel-Dieu, Paris, France

Acquired immunodeficiency syndrome (AIDS) is a retroviral infection caused by human immunodeficiency virus (HIV) [2, 17]. The disease is characterized by a variable incubation period in which the virus infects essentially cells with CD4 receptor, in particular immune cells [11]. Several points remain to be elucidated, such as the cofactors necessary to reactivate HIV from its latent form, and the viral dissemination in the body where specific cell-virus interactions could promote a particular type of pathology [10, 16].

To better understand the relationship between HIV and the infected cells, we analyzed by in situ hybridization the expression of the HIV genome in an in vitro lytic system of infection and in lymphoid tissue from an ARC lymph node. The cDNA probe specific for HIV was labelled with tritium for a more precise localization of the intracellular viral RNA.

Materials and Methods

Cells, Tissues and Virus

The human leukemic $CD4^+$ cell line H9 was grown at 10^6 cells/ml in RPMI 1640 medium supplemented with 10% fetal calf serum and 1,000 IU/μg/ml penicillin and streptomycin. The cell line was inoculated with the LAV strain of HIV virus at a concentration of 5,000 cpm units of reverse transcriptase activity per milliliter. After 10 days of infection, cells were processed for in situ hybridization.

[1]Supported by grants from INSERM (U. 43 and PNRS No. 3).

[2]The authors thank I. Tardivel for her technical assistance and L. Sergent and M. C. Rufino for their aid in the manuscript preparation.

An AIDS-related complex lymph node was used. Histological observations showed a follicular hyperplasia with numerous free retroviral particles detected by electron microscopy [see J. Diebold et al., present review].

Probes

The pBT1 probe containing 9 kb of HIV genome in the plasmid pUC 18 was used (a generous gift from S. Wain-Hobson and R. Vazeux, Institut Pasteur, Paris). The probe was radiolabelled by nick translation with ^{3}H-dCTP and ^{3}H-TTP to a specific activity of 1–2 10^{7} cpm/μg. The control probe pUC 18 was labelled in a similar manner.

In situ Hybridization

The specimens were frozen in liquid nitrogen and stored at –80°C. Five micrometer frozen sections collected on clean microscope slides were fixed in 4% buffered formaldehyde for 15 min, dehydrated, and stored until use. Cells were cytocentrifuged and, after a few minutes drying, fixed as described above.

Prior to hybridization, the sections were incubated with 0.2 *M* HCl for 10 min, then with 5 μg/ml proteinase K at +37°C for 15 min, and cells were digested by pronase at 0.01% for 2 min at room temperature [5, 6].

Hybridization was carried out under a sealed coverslip at +37°Cfor 20 h with 0.1 μg/ml DNA probe in a solution containing 50% freshly deionized formamide, 10% dextran sulfate, 600 m*M* NaCl, 10 m*M* Tris HCl, 1 m*M* EDTA, 0.02% each of ficoll, povidone and BSA, 1 mg/ml *Escherichia coli* and yeast RNA, and 400 μg/ml herring and salmon sperm DNA.

After hybridization, slides were washed once in 50% formamide 4 × SSC for 1 min, once in 50% formamide 2 × SSC for 1 h, twice in 2 × SSC for 30 min, and once in 1 × SSC for 30 min.

After dehydration, slides were coated with Kodak NTB2 emulsion, exposed at +4°C for 4–7 days (cells) and 1–2 months (tissues), developed in Kodak D19, and stained with MGG.

Results

In infected H9 cells, the HIV probe hybridized with numerous cells (30–50%). Individual cells but also cells fusioned in polykaryonic formations showed a hybridization signal corresponding to viral HIV-RNA (fig. 1a), since no specific HIV signal was seen in noninfected cells or infected cells hybridized with the pUC 18 control plasmid (fig. 2b). Among the positive cells, many showed a peripheral distribution of viral RNA close to the plasma membrane (fig. 1b, c), certain of them with a polar localization (fig. 1c, d). This cytoplasmic presence of viral RNA was accompanied or not by the presence of nuclear RNA. In the lymph node from a patient with an AIDS-related complex, positive cells were observed in lymph follicles, either in the germinal centers or in the mantle zone. Although the morphological identification of the cells after ISH on cryosections was difficult (fig. 3a), the majority of the radioactive

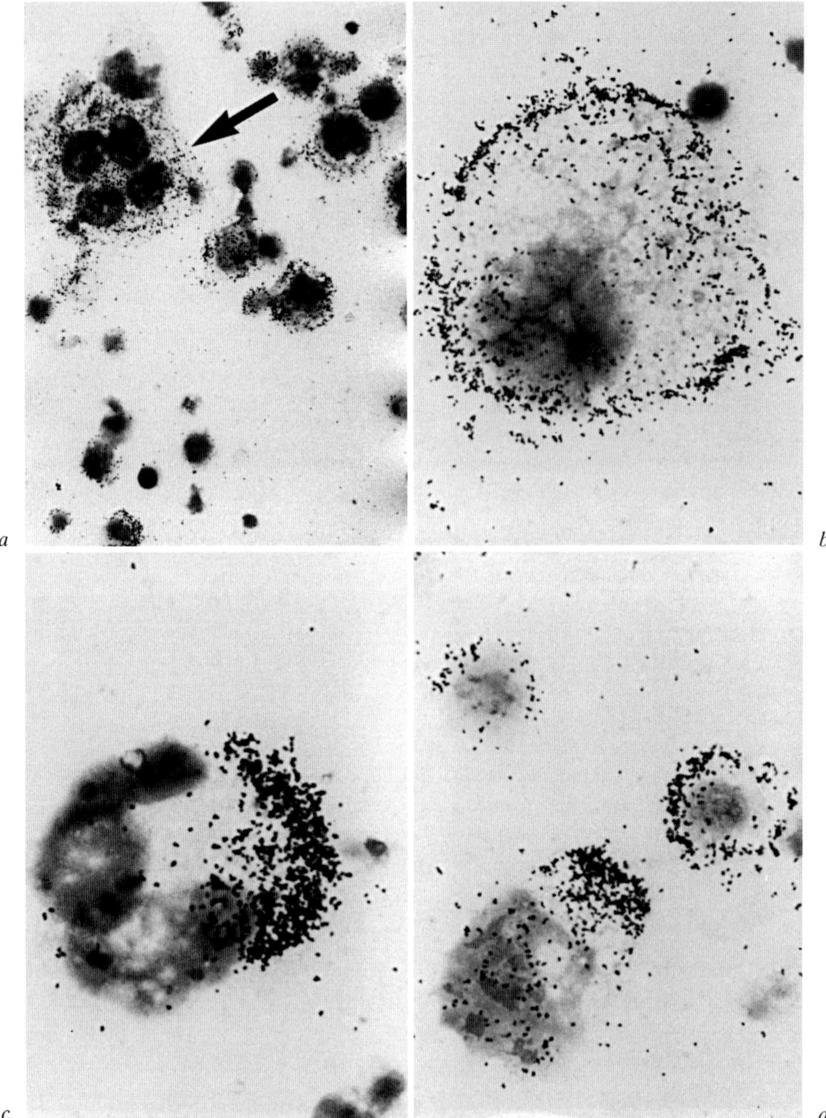

Fig. 1. Infected HIV-H9 cells hybridized with the tritiated pBT_1 probe specific for HIV. *a* Individual and polykaryonic cells (arrow) containing HIV-RNA molecules. *b* Peripheral localization of the HIV-RNA. *c, d* Polar distribution of the cytoplasmic HIV-RNA.

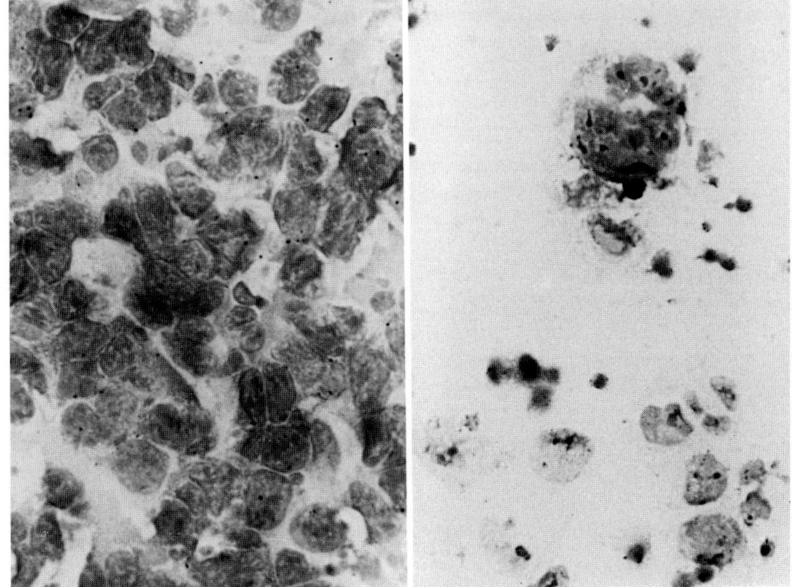

Fig. 2. ARC lymph node *(a)* and HIV-infected H9 cells *(b)* hybridized with the pUC 18 control probe. Only background can be seen.

cells seem to be of the lymphocyte type. In addition to the HIV-specific signal in distinct cells, a more diffuse signal was seen exclusively in most of the germinal centers (fig. 3b).

Discussion

HIV genome expression was analyzed at the cellular level by the in situ hybridization technique, using a cDNA probe containing 9 kb of the HIV genome. With in vitro infected HIV-H9 cells, the presence of HIV-RNA was observed after 10 days of infection in 30–50% of cells, some of which were included in polykaryonic formations illustrating the fusion property of the HIV virus [12]. In HIV-positive cells, a peripheral localization of the viral RNA was seen in the majority of cells. Such precise detection was possible because the probe was labelled after nick translation with tritium, known to provide the best resolution of the autoradiographic signal [18]. This local-

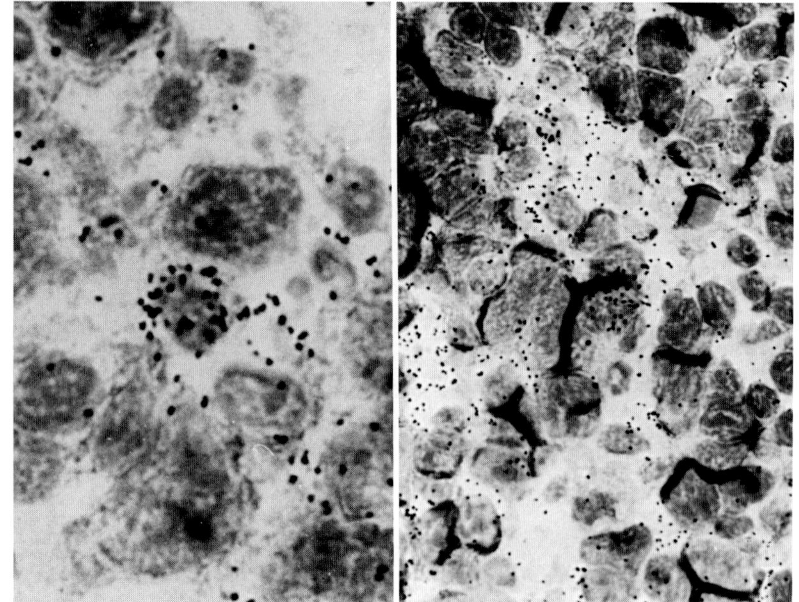

Fig. 3. ARC lymph node tissue cryosections hybridized with the pBT1 probe. *a* Individual HIV-positive cells with the appearance of lymphocytes present in the germinal center. *b* Diffuse hybridization signal in the germinal center.

ization might mean that the RNA molecules synthesized in the nucleus from the proviral DNA are not accumulated in the cytoplasm but immediately transported to the periphery of the cell, where it is known that the maturation of the virions occurs at the plasma membrane level [10]. The rapid migration to the periphery of the structural components of the virion is also observed for proteins, since several ultrastructural immunocytochemistry studies have indicated that the viral antigens are never detected in the cytoplasm but always in the budding processes or in the extracellular mature particles [8, 9, 14]. Furthermore, the detection of the accumulation of HIV-RNA in part of the cell periphery is in agreement with the production of the virus at a preferred site of the plasma membrane, clearly observable using scanning electron microscopy [3], which indicates that the morphogenesis of the virus could be carried out in a polar manner in the infected cell. In our experimental conditions, one cannot however consider that the genomic viral RNA is detected in the mature virion, but only the genomic and messenger RNAs free in the cytoplasm. Indeed, several ultrastructural ISH studies have shown that in cells infected with papilloma virus [4], SV40 [7] or CMV [20], the viral nucleic acids present within the capsid are never or hardly ever hybridized with the probe, whereas the viral nucleic acids free in the nucleoplasm are. Nevertheless, the peripheral detection of HIV-RNA in infected cells could be a marker of active virion production.

The in vivo detection of cells infected with HIV was performed on a lymph node from a patient with an AIDS-related complex. In this tissue the main cell type apparently containing HIV-RNA is the lymphocyte present in the germinal center and in the mantle zone of the follicule [3, 15, 19]. In addition, a diffuse autoradiographic signal is observed in most of the germinal centers. This hybridization signal has been attributed to the extracellular virus particles visible between the FDC processes using electron microscopy [3]. However, based on the above interpretations concerning in vitro infected cells, it seems likely that the diffuse signal could also reflect the presence of intracytoplasmic viral RNA in close proximity to the plasma membrane of the FDC. This would imply, as suggested by several authors [1, 3, 19, and S. Prevot et al., Pathol. Res. Pract., in press], that the FDC is infected with HIV and actively produces mature viral particles.

References

1 Armstrong, J. A.; Horne, R.: Follicular dendritic cells and virus-like particles in AIDS-related lymphadenopathy. Lancet *i:* 370–372 (1985).

2 Barré-Sinoussi, E.; Chermann, J. C.; Rey, F.; Nugeyre, M. T.; Chamaret, S.; Gruest, J.; Dauget, C.; Axler-Blin, C.; Vezinet-Brun, F.; Rouzioux, C.; Rozenbaum, W.; Montagnier, L.: Isolation of a T-lymphotropic retrovirus from a patient at risk for acquired immunodeficiency syndrome (AIDS). Science *220:* 868–871 (1983).

3 Biberfeld, P.; Chayt, K. J.; Marselle, L. M.; Biberfeld, G.; Gallo, R.; Harper, E.: HTLV-III expression in infected lymphnodes and relevance to pathogenesis of lymphadenopathy. Am. J. Path. *125:* 436–442 (1986).

4 Croissant, O.; Dauguet, C.; Jeanteur, P.; Orth, G.: Application de la technique d'hybridation moleculaire in situ à la mise en évidence au microscope électronique, de la réplication végétative de l'ADN viral dans les papillomes provoqués par le virus de Shope chez le lapin cottontail. C. r. hebd. Séanc. Acad. Sci., Paris *274* (sér. D); 614–617 (1972).

5 Fournier, J. G.; Tardieu, M.; Lebon, P.; Robain, O.; Ponsot, G.; Rozenblatt, S.; Bouteille, M.: Detection of measles virus RNA in lymphocytes from peripheral blood and brain perivascular infiltrates of patients with subacute sclerosing panencephalitis. New Engl. J. Med. *313:* 910–915 (1985).

6 Fournier, J. G.; Lebon, P.; Bouteille, M.; Goutières, F.; Rozenblatt, S.: Subacute sclerosing panencephalitis: detection of measles virus RNA in appendix lymphoid tissue before clinical signs. Br. med. J. *293:* 523–524 (1986).

7 Geuskens, M.; May, E.: Ultrastructural localization of SV40 viral DNA in cells during lytic infection by in situ molecular hybridization. Expl Cell Res. *87:* 175–185 (1974).

8 Grigoriev, V. B.; Escaig, F.; Fournier, J. G.; Kadochnicov, J. P.; Rudneva, I.; Karamov, E.; Klimenko, S. M.: Detection by immunogold techniques of HIV antigens in Lowicryl ultrathin sections of infected cells. Pathol. Res. Pract. *184:* 494–497 (1989).

9 Hausmann, E. H. S.; Gelderblom, H. R.; Clapham, P. R.; Pauli, G.; Weiss, R. A.: Detection of HIV envelope specific antibodies by immuno-electron microscopy with antibody titer and neutralizing activity. J. virol. Meth. *16:* 125–137 (1987).

10 Ho, D. D.; Pomerantz, R. J.; Kaplan, J. C.: Pathogenesis of infection with human immunodeficiency virus. New Engl. J. Med. *317:* 278–286 (1987).

11 Klatzmann, D.; Barre-Sinoussi, F.; Nugeyre, M. T.; Dauguet, C.; Vilmer, E.; Griscelli, C.; Brun-Vezinet, F.; Rouzioux, C.; Gluckman, J. C.; Montagnier, L.: Selective tropism of lymphadenopathy-associated virus (LAV) for helper-inducer T lymphocytes. Science *225:* 59–63 (1984).

12 Lifson, J. D.; Coutré, S.; Huang, E.; Engelman, E.: Role of envelope glycoprotein carbohydrate in human immunodeficiency virus (HIV) infectivity and virus induced cell fusion. J. exp. Med. *164:* 2101–2106 (1986).

13 Pautrat, G.; Morel, G.; Dihl, F.; Allasias, C.; Kourilsky, F.; Chermann, J. C.: Polar distribution of viral particles on the cell surface of a lymphoid cell line in vitro infected with the virus HIV intracellular. Retrovirus *1:* 17–24 (1988).

14 Pekovic, D.; Garzon, S.; Strykowski, H.; Ajdukovic, D.; Gornitsky, M.; Dupuy, J. M.: Immunogold labeling of HTLV-III/LAV in H9 cells studied by transmission electron microscopy. J. virol. Meth. *13:* 265–269 (1986).

15 Pezella, M.; Pezzella, F.; Galli, C.; Macchi, B.; Verani, P.; Sorice, F.; Baroni, C. D.: In situ hybridization of human immunodeficiency virus (HTLV-III) in cryostat sections of lymph nodes of lymphadenopathy syndrome patients. J. med. Virol. *22:* 135–142 (1987).

16 Pomerantz, R. J.; Kuritzkes, D. R.; de la Monte, S. M.; Rota, T. R.; Baker, A. S.; Albert, D.; Bor, D. H.; Feldman, E. L.; Schooley, R. T.; Hirsch, M. S.: Infection of the retina by human immunodeficiency virus type I. New Engl. J. Med. *317:* 1643–1647 (1987).

17 Popovic, M.; Sarngadharan, M. G.; Read, E.; Gallo, R.: Detection, isolation and continuous production of cytopathic retroviruses (HTLV-III) from patients with AIDS and pre-AIDS. Science *224:* 497–500 (1984).

18 Rogers, A. W.: Techniques of autoradiography; 3rd ed. (Elsevier/North-Holland Biomedical Press, Amsterdam 1979).

19 Tenner-Racz, K.; Racz, P.; Gluckman, J. C.; Popovic, M.: Cell-free HIV in lymph nodes of patients with AIDS and generalized lymphadenopathy. New Engl. J. Med. *318:* 49–50 (1988).

20 Wolber, R. A.; Beals, T. F.; Lloyd, R. V.; Maassab, H. F.: Ultrastructural localization of viral nucleic acid by in situ hybridization. Lab. Invest. *59:* 144–148 (1988).

J. G. Fournier, INSERM U. 43, Hôpital St. Vincent de Paul,
74, Av. Denfert-Rochereau, F–75674 Paris Cédex 14 (France)

Racz P, Haase AT, Gluckman JC (eds): Modern Pathology of AIDS and Other Retroviral Infections. Basel, Karger, 1990, pp 69–81

HIV Replication in the Central Nervous System of AIDS Patients

R. Vazeux, M.C. Cumont, L. Montagnier

Unité d'Oncologie virale et UA CNRS 04 1157, Institut Pasteur, Paris, France

The human immunodeficiency virus (HIV) induces subacute encephalitis with severe dementia in about 20% of adult acquired immunodeficiency syndrome (AIDS) cases [23, 27, 32]. HIV has been conclusively demonstrated in the central nervous system (CNS) only in monocytes/macrophages/microglial cells [18, 29, 38]. Infection of other cell types is controversial [9, 12, 30, 35, 39, 41]. Infected cells in the CNS express high levels of viral transcripts and viral antigens in immunosuppressed patients. One mechanism to explain the extensive HIV replication observed in the CNS of some patients, but not in others, might be the association of additional opportunistic infections. Associated pathogens might trigger or enhance HIV replication in CNS either by release of cytokines from infected cells, recruitment and activation of brain macrophages or by synergism of the two viruses in the same cell. In this paper we will discuss the CNS cell types infected by HIV, the mechanism of encephalopathy and will explore the association between HIV encephalitis and additional opportunistic infections using cases of adult HIV/JC virus CNS coinfections as a model.

Progressive multifocal leukoencephalopathy (PML) is caused by a polyomavirus, the JC virus, which replicates in CNS mainly in oligodendrocytes [1, 15, 40] and induces a demyelinating encephalitis in about 3–5% of AIDS patients [2, 17, 21, 24, 42]. We have extensively screened frozen CNS samples of 85 AIDS patients for the presence of HIV and associated pathogens, and we have observed 2 cases of HIV extensive encephalitis associated with PML. The use of double immunohistological staining and combination of immunohistochemistry with in situ hybridization were especially powerful to investigate the viral replication in CNS and the relationship between the HIV infection and an associated opportunistic infection.

Neuropathological Study

CNS tissue blocks were obtained at autopsy from 85 adult AIDS patients (often selected for neurological symptoms), 2 uninfected controls, and 1 patient with PML who was not infected with HIV. For each patient we studied five different areas: hemispheres (frontal, parietal, temporal cortex), basal ganglia, brain stem, cerebellum, and when possible, spinal cord. To preserve cellular and viral antigens as much as possible, each tissue block was frozen in isopentane cooled in liquid nitrogen, subdivided and stored at –80°C.

Neuropathological examination on 10% formalin-fixed tissues was performed in all cases. Gross examination was performed on coronal sections, and paraffin or celloidin-embedded tissues were examined after usual stainings (hematoxylin and eosin, Bodian, Luxol fast blue, Luxol fast blue cresyl violet, etc.).

Histochemistry

Tissue sections were deposited on slides coated with neoprene, fixed in acetone, and stored at –20°C for immunological staining or in situ hybridization. We studied at least 15–20 tissue blocks for each patient.

Each CNS sample was checked for the presence of HIV antigens using monoclonal antibodies directed against the p18, p24 and against the gp41 and gp110 viral antigens. Polyomavirus were identified with rabbit polyclonal antibodies directed against the total disrupted SV40 and against the large T antigen of SV40 [36, 37]. CNS samples were checked for the presence of associated infections with antibodies directed against toxoplasma antigens, cytomegalovirus early and late antigens, herpes simplex 1 and 2 antigens (table 1). CNS and inflammatory cells were characterized with a panel of antibodies directed against specific cellular antigens (table 2).

Monoclonal primary antibodies were labeled with a three-stage indirect immunoperoxidase technique [34] and primary polyclonal rabbit antibodies with a peroxidase-antiperoxidase technique (Dakopatts). They were used in single or sequential double immunological stainings according to published procedures [25, 38]. All stainings were performed after incubation with 3–75% normal human serum (diluted in Tris-buffered saline, pH 7.4) and 1% bovine serum albumin to avoid nonspecific binding. This is especially important in CNS tissues presenting astrocytosis and reactive microglia.

Table 1. Antibodies directed against HIV, SV40, CMV, toxoplasma, and HSV antigens

Antibody	Source	Specificity
Anti-p18	Genetic systems	internal HIV protein
Anti-p24	Genetic systems Epitope, Inc. Biosoft/Clonatec	internal HIV protein
Anti-gp41	Genetic systems	surface HIV glycoprotein
Anti-gp110	Genetic systems	surface HIV glycoprotein
Anti-SV40	E. May (CNRS, UPR 4)	late SV4O antigen
Anti-SV40	Lee Biomolecular	late SV4O antigen
Anti-T-SV40	Oncogene Science, Inc.	early SV4O antigen
Anti-CMV-E 13	Biosoft/Clonatec	early CMV antigen
Anti-CMV	Michelson et al. [26]	early CMV antigen
Anti-CMV	J. C. Nicolas (Hôpital Trousseau)	late CMV antigen
Anti-toxoplasma-G2	Biosoft/Clonatec	toxoplasma antigen
Anti-HSV1/HSV2-CHA 437	Biosoft/Clonatec	common HSV antigen

In situ Hybridization

Hybridization techniques, RNA labelling and combined hybridization with immunohistological staining were based upon modifications of published procedures [4, 7, 31]. Sections fixed in acetone and stored at -20°C were dried and fixed in paraformaldehyde 4%, then acetylated in 0.1 *M* triethanolamine solution pH 8, 0.25% anhydride acetic for 10 min, rinsed and dehydrated. Probes were then applied directly onto the tissue section to a final concentration of 50,000 cpm/μl in hybridization buffer (50% formamide, 10% dextran sulfate, 0.3 *M* NaCl, 20 m*M* Tris pH 7.5, 5 m*M* EDTA, 1 × Denhart, 0.5 mg/ml yeast RNA, 10 mMDTT), covered with siliconized coverslip and incubated in humid chamber at 42°C during 16 h. They were washed in 50% formamide, 2 × SSC, 10 m*M* DTT at 60°C for 30 min, twice in 1 × SSC, 0.1 × SSC at room temperature with agitation for 30 min and dehydrated. If background problems occurred, we added a second wash in 50% formamide, 2 × SSC, 10 m*M* DTT at 60°C for 30 min or RNAse A 20 μg/ml for 30 min. Exposure time for autoradiography ranged from 3 days to a few weeks. As controls, hybridization was performed on brain tissue sections of uninfected patients and patients with PML or HIV encephalitis.

Table 2. Antibodies directed against CNS and inflammatory cells

Antibody	Source	Specificity
Anti-GFAP	Dakopatts	astrocytes glial fibrillary acidic protein
Anti-neurofilaments	D. Paulin (Pasteur Institute)	neurofilaments
Anti-factor VIII	Dakopatts	endothelial cells
Leu M3	Becton-Dickinson (B-D)	monocytes
Kim7	Behring	macrophages/microglial cells
Leu M5	B-D	macrophages/microglial cells
Kim6	Behring	macrophages/phagocytes
CD4 Leu 3a	B-D	helper/inducer T cells, macrophages
CD3 Leu4	B-D	all T lymphocytes
CD8 Leu2a	B-D	suppressor/cytotoxic T cells
CD19 Pan B	B-D	B cells
Anti-HLA-DR	B-D	histocompatibility class II antigens
IOT6	Immunotech	interdigitating cells
DRC 1	Dakopatts	dendritic reticular cells

We combined immunohistological staining with in situ hybridization on the same tissue section. In a first step, frozen tissue sections were incubated in paraformaldehyde 4% for 20 min, rinsed in PBS then processed for immunohistological staining. Then, the same tissue sections were fixed in paraformaldehyde 4% for 10 min, acetylated and hybridized as previously described.

Results

Massive HIV infection was observed in the frozen CNS tissues of 24/89 patients. In those patients, histological features of subacute encephalitis were detected on formalin-fixed tissues, associated with foci of demyelination, inflammatory infiltrates, disseminated microglial nodules and multinucleated cells. No vacuolation or histological lesions could be observed in the spinal cords. A lower degree of infection was detected in an additional 11 patients with only 1 or 2 clusters of HIV-infected cells and without subacute encephalitis. We could not detect any infected cells in 49 patients.

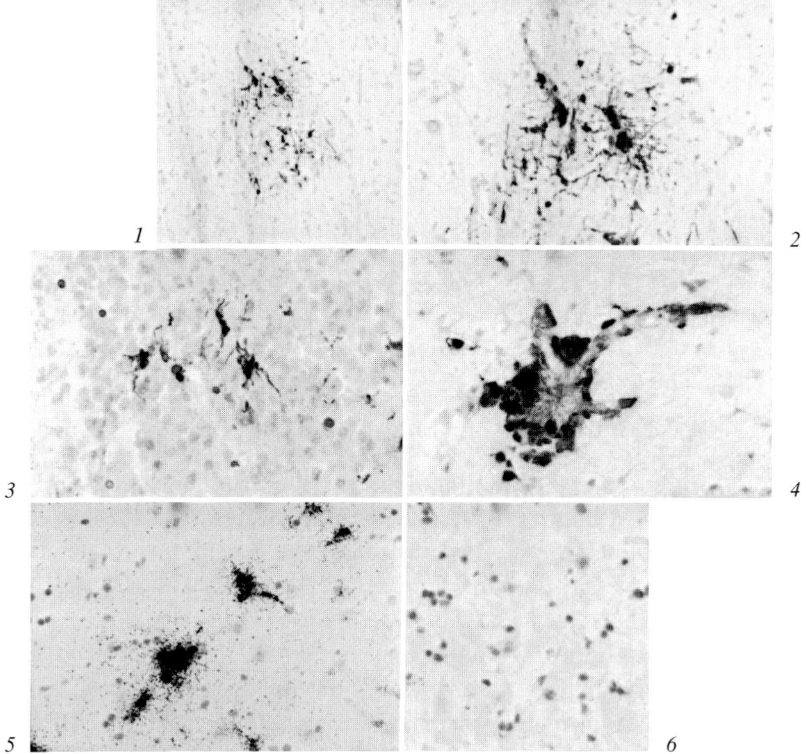

Fig. 1–4. Immunohistochemical staining for viral antigens. *1* Cluster of HIV-infected cells in hemispheres (white matter), labeled with anti-gp41 monoclonal antibody. ×250. *2* Higher magnification of the precedent slide. ×400. *3* Mononucleated cells scattered in hippocampal formation, labeled with p 25. ×400. *4* Perivascular infiltration by multinucleated cells, labeled with p 25. ×400.

Fig. 5, 6. 5 HIV-infected cells detected by RNA in situ hybridization after 5 days of exposure, in white matter of cerebral hemisphere. Autoradiography and HE. ×400. *6* Control brain hybridized with the same probe and exposed the same time.

In the 24 patients with massive infection, clusters of HIV-infected cells were detected in almost every tissue block studied. In virtually all cases, both internal antigens, surface glycoproteins, and HIV RNA were expressed in infected cells (fig. 1–6). Infected cells predominated in basal ganglia, brain stem, white matter or grey and white matter junction. They were also detected in thoracic spinal cord, in posterior columns, and posterior roots. In the most extensive cases of encephalitis, infected cells were disseminated in parenchyma (fig. 3, 5), while in the mildest ones they were mostly perivascular. Multinucleated cells were observed in the most inflammatory cases, often located in perivascular infiltrates and microglial nodules (fig. 4). However, some extensive HIV infections could be detected by immunological staining even in cases with little inflammation and few multinucleated cells.

In 2 patients, HIV encephalitis was especially extensive with a very unusual aspect of infected cells. Multiple foci of demyelination were detected, with enlarged oligodendrocytes stained on frozen sections with antibodies directed against SV40 (fig. 7, 8) and labeled by in situ hybridization with JC probes (fig. 9, 10). These results demonstrated that these 2 patients had a PML related to the JC virus. HIV-infected cells were present in huge numbers in both patients, inside and outside demyelination areas, with relatively little inflammation and few multinucleated cells. Inside the demyelinated areas, infected cells had a very unusual large and foamy cytoplasm without processes (fig. 11), while outside demyelination areas they had the usual aspect of mononucleated with thin cytoplasm and long processes (fig. 12), sometimes associated in both areas with multinucleated cells. Except for a few cells that expressed early cytomegalovirus antigens in the brain stem of patient 1, no other pathogen could be identified in these 2 patients.

All cells expressing HIV antigens were $GFAP^-$ (fig. 13), $neurofilament^-$ (fig. 14), factor VIII^-. All were labeled with macrophage markers (fig. 15), and had the following phenotype: $Kim7^+$, $Kim6^+$, $HLA\text{-}DR^+$, $T6^-$, DRC^-. In addition, the majority of them were Leu $M3^-$, $CD4^-$. Others, mainly in perivascular areas, were Leu $M3^+$, $CD3^{+/-}$. When present in inflammatory infiltrates and microglial nodules, infected macrophages were Leu $M3^+$, $CD4^-$ or Leu $M3^+$, $CD4^+$. A few infected $CD3^+/CD8^-$ cells, presumably $CD4^+$ T lymphocytes, were also observed in these infiltrates. Thus the concurrent JC infection did not alter the exclusive macrophage/microglial tropism of HIV infection. In contrast, polyomavirus antigens were not present in macrophages, endothelial cells, neurons and few astrocytes were stained. We could not detect any cells expressing both HIV and JC antigens (fig. 16) and by combination of immunological staining for HIV antigens and RNA in situ hybridization with JC probe, we were unable to detect any double infection in the same cells (fig. 17).

Uninfected macrophages were present both in AIDS and control brains. They showed the same aspect as infected ones and had the following phenotype: Leu $M3^-$, $Kim7^+$, $Kim6^{+/-}$, $HLA\text{-}DR^+$, $T6^-$, DRC^-, $CD4^{+/-}$. In AIDS brains, present in inflammatory infiltrates, uninfected macrophages were Leu $M3^+$, $Kim7^+$, $Kim6^{+/-}$, $HLA\text{-}DR^+$, $T6^-$, DRC^-, $CD4^+$. This type of cell was not

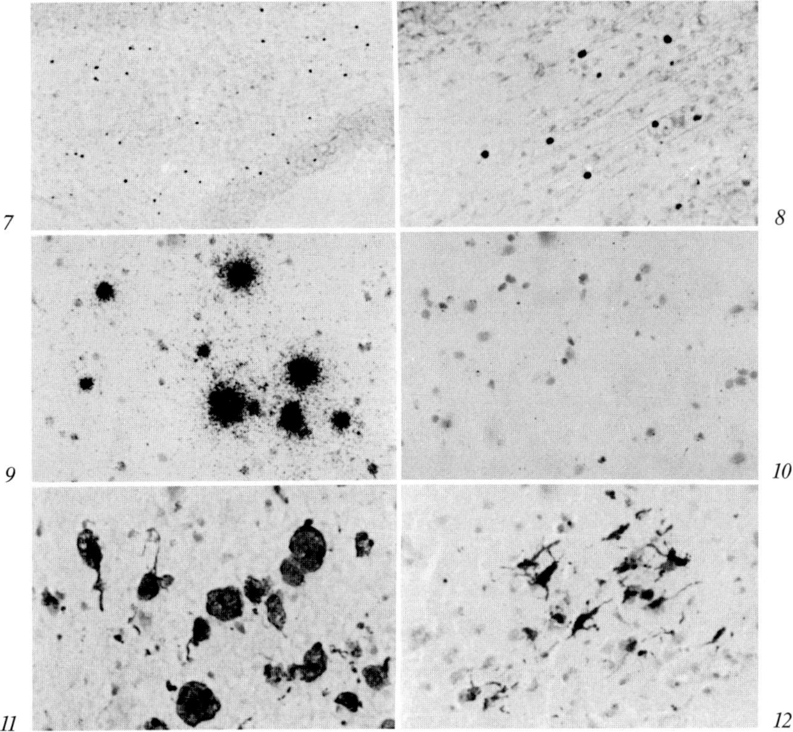

Fig. 7, 8. 7 Demyelinated area in the cerebellar white matter, surrounded by infected cells labeled with anti-SV40 polyclonal antibody. ×250. *8* Higher magnification of the precedent slide. ×400.

Fig. 9, 10. 9 JC-infected cells detected by RNA in situ hybridization after 2 days of exposure, in a demyelinated area. Autoradiography and HE. ×400. *10* Control brain hybridized with the same probe and exposed the same time.

Fig. 11, 12. Variable aspect of HIV-infected cells in CNS of patients with PML *11* Enlarged infected mononucleated cells inside a demyelinated area labelled with anti-p18 antibody. ×400. *12* Mononucleated cells disseminated in brain stem of the same patient, labeled with anti-p25, outside a demyelination area. ×400.

found in controls. 'Microglial nodules' were formed of inflammatory cells, mostly Leu $M3^+/CD4^+$ macrophages, $CD8^+$ and $CD3^+/CD8^-$ T lymphocytes and a small number of B cells. T lymphocytes ($CD8^+$ and $CD3^+/CD8^-$) were also scattered in grey and white matter parenchyma. $CD4^+$ T lymphocytes were not observed.

Discussion

The high rate of HIV encephalitis that we observed may be explained by the selection of several patients for neurological symptoms. HIV severe replicative encephalitis rate is certainly close to 15–20% dementia reported in large series of unselected AIDS patients [23, 32].

The appearance of replication in the CNS of only 20% of AIDS patients may be explained by several causes. First, several groups report a variable tropism of HIV in relation to different strains. Some HIV strains replicate preferentially in macrophages while other replicate exclusively in CD4 lymphocytes [6, 10, 19]. HIV strains could be neurotropic when specially targeting macrophages.

Second, there is a close relationship between the appearance of HIV encephalopathy and severity of immunosuppression [29]. Amongst HIV-infected individuals, dementia has been described only in AIDS patients where it may be the first and only symptom [28]. The pathogenesis of other neurological symptoms described in non-immunosuppressed patients or in ARC and AIDS are unknown without any evidence for HIV replication in CNS [5, 13, 16]. Therefore, the neurovirulence of HIV is caused by immunosuppression induced by HIV replication in CD4 lymphocytes.

Finally, a possible mechanism to explain development of HIV encephalitis is the triggering or enhancement of HIV replication in CNS by other pathogens with which it is very often associated [24]. PML and HIV encephalitis were good models to study such viral interactions since both viral infections were closely intermingled in 2 patients with demyelination areas induced by JC being massively invaded by HIV-infected cells. In those 2 cases, the extension of HIV encephalitis was extremely severe with the presence of huge numbers of HIV-infected cells that were disseminated in parenchyma and perivascular areas. Enlarged macrophages in demyelinated areas were either HIV-infected or uninfected. They are not specific for PML and could be observed in necrotic areas in other patients. Thus, these macrophages seem recruited to phagocyte-damaged tissues. As oligodendrocytes infection by JC virus induces large areas of white matter lesions and since HIV replication may be triggered when infected monocytes/macrophages are activated [8, 20], the most obvious mechanism that explains the especially impressive HIV encephalitis observed in those patients is the local recruitment and activation of monocytes/macrophages/microglial cells in response to CNS injury induced by JC virus.

Although an increase of HIV replication has been demonstrated in vitro

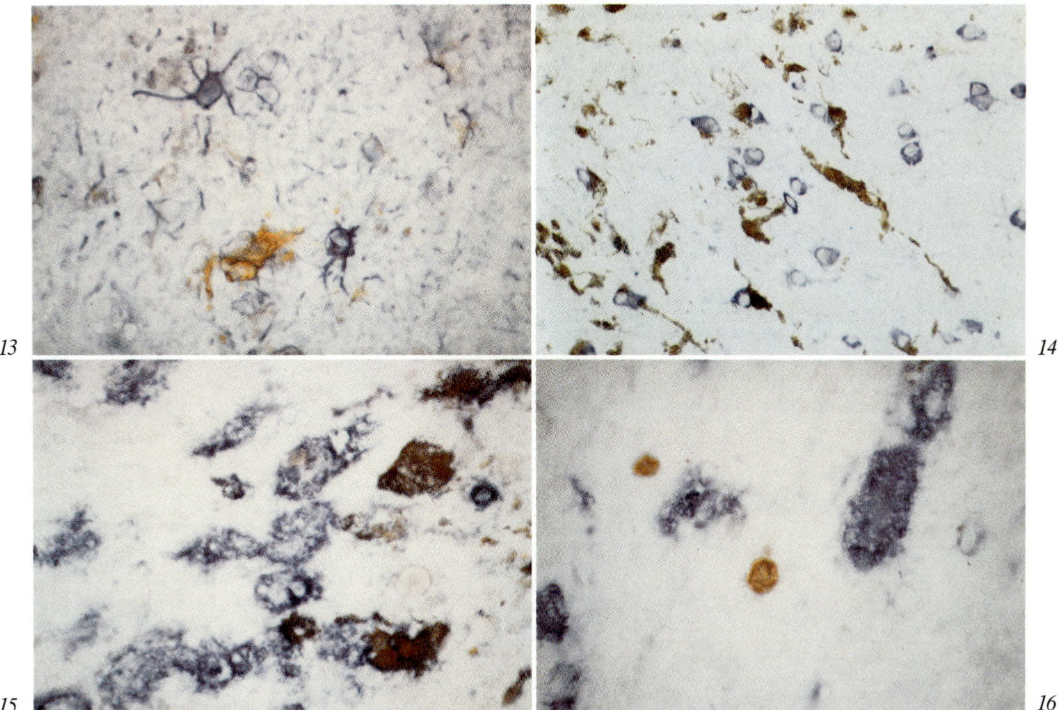

Fig. 13–15. Identification of infected cells by peroxidase/alkaline phosphatase double immunostaining, for viral p25 (brown) and cellular antigens (blue). ×245. *13* Astrocytes (GFAP, blue) are uninfected and process-bearing cells ($p25^+$) are $GFAP^-$. *14* Neurons (neurofilament, blue) are uninfected and HIV-infected cells ($p25^+$) are negative for neuronal markers. *15* Infected enlarged cells in demyelinated area are labeled for viral and macrophage markers (Kim7). Note the presence of associated uninfected macrophages.

Fig. 16. HIV- and JC-infected cells associated in demyelination area. HIV-infected cells express only HIV markers (p25) while JC-infected cells are labeled only with anti-SV40 antibody. ×245.

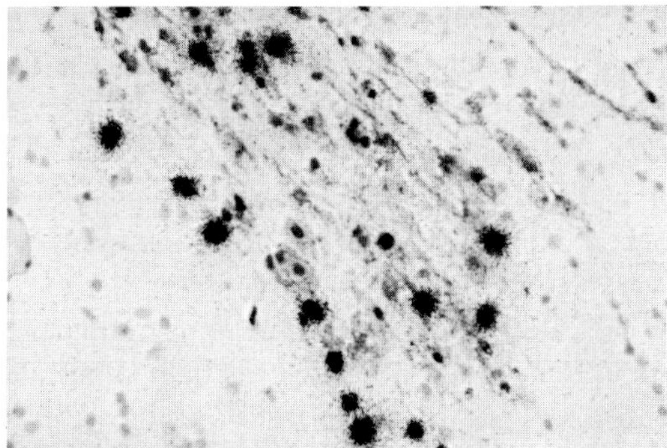

Fig. 17. Combination of immunological staining for HIV antigens (p25) and RNA in situ hybridization with JC probe. JC-infected cells surrounded HIV-infected macrophages without double infection in the same cell. Autoradiography and hematoxylin. × 400.

in cells *transfected* with both HIV and JC plasmids [11], we were unable to demonstrate the presence of both viruses in the same cell by double detection of HIV and JC antigens or by searching for expression of JC RNA in HIV-replicating cells. Thus, the transactivation obtained in vitro in artificial systems is not necessarily relevant to mechanisms observed in vivo that require *entry* of both viruses in the same cell. It is, therefore, very unlikely that this mechanism is a major factor for increasing HIV CNS infection.

As we could not detect any infected astrocytes, or neurons, and since HIV could not be detected in JC-infected cells, it is very unlikely that oligodendrocytes were infected. A major question in AIDS-related dementia is how neurological symptoms are induced by the virus that is located only in monocytes/macrophages/microglial cells. One possibility is that neurological symptoms might be explained by HIV latent infection in neurons or glial cells. More likely, there is an unfavorable interaction between CNS cells and HIV-infected macrophages, either by suppression of normal microglial function or by release of toxic products [22].

After it was discovered that HIV belongs to the lentivirus subfamily of retroviruses, a parallel was drawn between AIDS encephalopathy and visna, the prototype lentiviral infection [33]. Both viruses induce neurological disease but study of viral replication and identification of infected cells has revealed striking differences in the behavior of the two viruses in CNS. Visna induces slow demyelinating disease of sheep evolving over years. In visna encephalitis, the agent is present primarily in monocytes but also in astrocytes, oligodendrocytes and choroid plexus and viral replication in these cells is restricted at the transcriptional level [3, 14]. In AIDS patients HIV encephalitis evolves over weeks or months, and is associated with active transcription and translation of the viral genome. Recently, minor neuropsychiatric symptoms have been described in nearly 60% of ARC, AIDS patients and even in asymptomatic carriers [13, 16]. Therefore, it is possible that other encephalopathies exist in HIV-infected patients which resemble visna more closely.

In conclusion, a common feature for lentiviruses is their replication in macrophages and their neurotropism. Association with immunosuppression in HIV-infected patients may probably explain many differences in behavior of HIV in relation to other lentiviruses. A nonspecific mechanism of macrophage recruitment and activation certainly plays a major role in HIV brain dissemination, it may be a general mechanism that increases the level of HIV replication in the brain of adult AIDS patients.

Acknowledgements

The authors wish to acknowledge our collaborators for providing valuable patient material. We are particularly grateful to Dr. C. Katlama, Prof. D. Henin and Prof. J. J. Hauw from Hôpital Pitie-Salpétrière, Paris; P. M. Girard, Prof. Saimot, C. Debroeker, Dr. C. Marche from Hôpital Claude-Bernard, Paris; Prof. C. Vedrennes from Hôpital Sainte-Anne, Paris; Dr. P. Sansonnetti, X. Nassif, C. Le Vilain and P. Trotot from Hôpital Pasteur, Paris. Dr. N. Brousse and A. Jarry from Hôpital Beaujon, Paris. We are very grateful for the helpful discussions and advice of M. Emerman and M. Brahic.

References

1 Aksamit, A. J.; Mourrain, P.; Sever, J. L.; Major, E. O.: Progressive multifocal leukoencephalopathy: Investigation of three cases using in situ hybridization with JC virus biotinylated DNA probe. Ann. Neurol. *18:* 490–496 (1985).

2 Berger, J. R.; Kaszovitz, B.; Donovan Post, J. M.; Dickinson, G.: Progressive multifocal leucoencephalopathy associated with human immunodeficiency virus infection. Ann. Intern. Med. *107:* 78–87 (1987).

3 Brahic, M.; Stowring, L.; Ventura, P.; Haase, A. T.: Gene expression in visna virus infection in sheep. Nature *292:* 240–242 (1981).

4 Brahic, M.; Haase, A. T.; Cash, E.: Simultaneous detection of viral RNA and antigens. Proc. Natl. Acad. Sci. USA *81:* 5445–5448 (1984).

5 Carne, C. A.; Smith, A.; Elkington, S. G.; Preston, F. E.; Tedder, R. S.; Sutherland, S.; Daly, H. M.; Craske, J.: Acute encephalopathy coincident with seroconversion for anti-HTLV III. Lancet *ii:* 1206–1208 (1985).

6 Cheng-Mayer, C.; Levy, J. A.: Distinct biological and serological properties of human immunodeficiency viruses from the brain. Ann. Neurol. *23:*suppl., pp. S58–S61 (1988).
7 Cox, K. H.; Deleon, D. V.; Angerer, L. M.; Angerer, R. C.: Detection of mRNAs in sea urchin embryos by in situ hybridization using asymmetric RNA probes. Dev. Biol. *101:* 485–502 (1984).
8 Folks, T. M.; Justement, J.; Kinter, A.; Dinarello, C. A.; Fauci, A.: Cytokine-induced expression of HIV-1 in a chronically infected promonocyte cell line. Science *238:* 800–802 (1987).
9 Gabudza, D. H.; Ho, D. D.; De la Monte, S. M.; Hirsch, M. S.; Rota, T. R.; Sobel, R. A.: Immunohistochemical identification of HTLV III antigen in brain of patients with AIDS. Ann. Neurol. *20:* 289–295 (1986).
10 Gartner, S.; Markovits, P.; Markovitz, D. M.; Kaplan, M. H.; Gallo, R.; Popovic, M.: The role of mononuclear phagocytes in HTLV III/LAV infection. Science *233:* 215–219 (1986).
11 Gendleman, H. E.; Phelps, W.; Feigenbaum, L.; Ostrove, J. M.; Adachi, A.; Howley, P. M.; Khoury, G.; Ginsberg, H. S.; Martin, M. A.: Transactivation of the human immunodeficiency virus long terminal repeat sequence by DNA viruses. Proc. Natl. Acad. Sci. USA *83:* 9759–9763 (1986).
12 Giorkey, F.; Melnick, T. L.; Gyorkey, P.: Human immunodeficiency virus in brain biopsies of patients with AIDS and progressive encephalopathy. J. Infect. Dis. *155:* 870–876 (1987).
13 Grant, I.; Atkinson, J. H.; Hesselink, J. R.; Kennedy, C.; Richman, D. D.; Spector, S. A.; McCutchan, J. A.: Evidence for early central nervous system involvement in the acquired immunodeficiency syndrome (AIDS) and other human immunodeficiency virus (HIV) infections. Ann. Intern. Med. *107:* 828–836 (1987).
14 Haase, A. T.: Pathogenesis of lentivirus infection. Nature *322:* 130–136 (1986).
15 Houff, S. A.; Major, E. O.; Katz, D. A.; Kufta, C. V.; Sever, J. L.; Pittaluga, S.; Roberts, J. R.; Gitt, J.; Saini, N.; Lux, W.: Involvement of JC virus-infected mononuclear cells from the bone-marrow and spleen in the pathogenesis of progressive multifocal leukoencephalopathy. N. Engl. J. Med. *318:* 301–305 (1988).
16 Janssen, R. S.; Saykin, A. J.; Kaplan, J. E.; Spira, T. J.; Pinsky, P. F.; Sprehn, G. C.; Hoffman, J. C.; Mayer, B.; Schonberger, L. B.: Neurological complications of human immunodeficiency virus infection in patients with lymphadenopathy syndrome. Ann. Neurol. *23:* 49–55 (1988).
17 Kleihues, P.; Lang, W.; Burger, P. C.; Budka, H.; Vogt, M.; Maurer, R.; Lüthy, R.; Siegenthaler, W.: Progressive diffuse leukoencephalopathy in patients with acquired immunodeficiency syndrome (AIDS). Acta Neuropathol. (Berl.) *68:* 333–339 (1985).
18 Koenig, S.; Gendelman, H. E.; Orenstein, J. M.; Dal Canto, M. C.; Pezeshkpour, G. H.; Yungbluth, P.; Janotta, F.; Aksamit, A.; Martin, M.; Fauci, A. S.: Detection of AIDS virus in macrophages in brain tissue from AIDS patients with encephalopathy. Science *233:* 1089–1093 (1986).
19 Koyanagi, Y.; Miles, S.; Mitsuyasu, R. T.; Merril, J. E.; Vinters, H. V.; Chen, I. S.: Dual infection of the central nervous system by AIDS viruses with distinct cellular tropisms. Science *236:* 819–822 (1987).
20 Koyanagi, Y.; O'Brien, W. A.; Zhao, J. Q.; Golde, D. W.; Gasson, J. C.; Chen, I. S.: Cytokines alter production of HIV-1 from primary mononuclear phagocytes. Science *241:* 1673–1675 (1988).

21 Krupp, L. B.; Lipton, R. B.; Swerlow, M. L.; Leeds, N. E.; Llena, J.: Progressive multifocal leukoencephalopathy: clinical and radiological features. Ann. Neurol. *17:* 344–349 (1985).
22 Lee, M. R.; Ho, D. D.; Gurney, M. E.: Functional interaction and partial homology between human immunodeficiency virus and neuroleukin. Science *237:* 1047–1051 (1987).
23 Levy, R. M.; Bredesen, D. E.; Rosenblum, M. L.: Neurological manifestations of the acquired immunodeficiency syndrome (AIDS): experience at UCSF and review of the literature. J. Neurosurg. *62:* 475–495 (1985).
24 Levy, R. M.; Bredesen, D. E.; Rosenblum, M. L.: Opportunistic central nervous system pathology in patients with AIDS. Ann. Neurol. *23:* suppl., pp. S7–S12 (1988).
25 Mason, D. Y.; Abdulaziz, Z.; Falini, B.; Stein, H.: In double immunoenzymatic labeling; in Polak, J. M.; van Noorden, S. (eds): Immunocytochemistry: Pracitcal Applications in Pathology and Biology, pp. 113–118 (Wright, London, 1983).
26 Michelson, S.; Tardy-Panit, M.; Barzu, O.: Properties of a human cytomegalovirus-induced protein kinase. Virology *134:* 259–268 (1984).
27 Navia, B. A.; Jordan, B. D.; Price, R. W.: The AIDS dementia complex: Clinical features. Ann. Neurol. *19:* 517–524 (1986).
28 Navia, B. A.; Price, R.: The acquired immunodeficiency syndrome dementia complex as the presenting or sole manifestation of human immunodeficiency infection. Arch. Neurol. *44:* 65–69 (1987).
29 Price, R.; Brew, B.; Sidtis, J., et al.: The brain in AIDS: Central nervous system HIV infection and AIDS dementia complex. Science *239:* 586–592 (1988).
30 Pumarola-Sune, T.; Navia, B. A.; Cordon-Cardo, C.; Cho, E. S.; Price, R.: HIV antigen in the brain of patients with the AIDS dementia complex. Ann. Neurol. *21:* 490–496 (1986).
31 Singer, R. H.; Lawrence, J. B.; Rashtchian, R. N.: Toward a rapid and sensitive in situ hybridization methodology using isotopic and nonisotopic probes; in Valentino, Eberwine, Barchas (eds): In situ Hybridization. Application to Neurobiology, pp. 71–96 (Oxford University Press, New York 1987).
32 Snider, W. D.; Simpson, D. M.; Nielsen, S.; Gold, J. W. M.; Metroka, C. E.; Posner, J. B.: Neurological complications of acquired immune deficiency syndrome: analysis of 50 patients. Ann. Neurol. *14:* 403–418 (1983).
33 Sonigo, P.; Alizon, M.; Staskus, K.; Klatzmann, D.; Cole, S.; Danos, O.; Retzel, E.; Tiollais, P.; Haase, A.; Wain-Hobson, S.: Nucleotide sequence of the visna lentivirus: relationship to the AIDS virus. Cell *42:* 369–382 (1985).
34 Stein, H.; Gerdes, J.; Schwab, U.; Lemke, H.; Mason, D. Y.; Ziegler, A.; Schienle, W.; Diehl, V.: Identification of Hodgkin and Sternberg-Reed cells as a unique cell type derived from a newly-detected small-cell population. Int. J. Cancer *30:* 445–459 (1982).
35 Stoler, M. H.; Eskin, T. A.; Benn, S.; Angerer, R. C.; Angerer, L. M.: Human T-cell lymphotropic virus type III infection of the central nervous system. A preliminary in situ analysis. JAMA *256:* 2360–2364 (1986).
36 Stoner, G. L.; Ryschkewitsch, C. F.; Walker, D. L.; Webster, H. F.: JC papovavirus large tumor (T)-antigen expression in brain tissue of acquired immunodeficiency syndrome (AIDS) and non-AIDS patients with progressive multifocal leucoencephalopathy. Proc. Natl. Acad. Sci. USA *83:* 2271–2275 (1986).
37 Stoner, G. L.; Ryschkewitsch, C. F.; Walker, D. L.; Sofer, D.; Webster, H. F.: A mono-

clonal antibody to SV40 large T-antigen labels a nuclear antigen in IC virus-transformed cells and in progressive multifocal leucoencephalopathy (PML) brain infected with JV virus. J. Neuroimmunol. *17:* 331–345 (1988).

38 Vazeux, R.; Brousse, N.; Jarry, A.; Henin, D.; Marche, C.; Vedrenne, C.; Mikol, J.; Wolff, M.; Michon, C.; Rozenbaum, W.; Bureau, J. F.; Montagnier, L.; Brahic, M.: AIDS subacute encephalitis: Identification of HIV-infected cells. Am. J. Pathol. *126:* 403–410 (1987).

39 Ward, J. M.; O'Cleary, T. J.; Baskin, G. B.; Benveniste, R.; Harris, C.; Nara, D.; Rhodes, R. H.: Immunohistological localization of human and simian immunodeficiency viral antigens in fixed tissue sections. Am. J. Pathol. *127:* 199–205 (1987).

40 Walker, D. L.: Progressive multifocal leukoencephalopathy: an opportunistic infection of the central nervous system. Handbook of Clinical Neurology, vol. 34: Infections of the Nervous System (North-Holland, Amsterdam 1978).

41 Wiley, C. A.; Schrier, R. D.; Nelson, J. A.; Lampert, P. W.; Oldstone, M. B. A.: Cellular localization of human immunodeficiency virus infection within the brains of acquired immunodeficiency syndrome patients. Proc. Natl. Acad. Sci. USA *83:* 7089–7093 (1986).

42 Wiley, C. A.; Grafe, M.; Kennedy, C.; Nelson, J. A.: Human immunodeficiency virus (HIV) and JC virus in acquired immunodeficiency syndrome (AIDS) patients with progressive multifocal leukoencephalopathy. Acta Neuropathol. *76:* 338–346 (1988).

Rosemay Vazeux, MD, Unité d'Oncologie virale et UA CNRS 04 1157,
Institut Pasteur, 25–28, Rue du Docteur Roux, F-75724 Paris Cédex 15 (France)

Racz P, Haase AT, Gluckman JC (eds): Modern Pathology of AIDS and Other Retroviral Infections. Basel, Karger, 1990, pp 82–98

Some Comparative Aspects of Visna and AIDS

Gudmundur Georgsson, Páll A. Pálsson, Gudmundur Pétursson[1]

Institute for Experimental Pathology, University of Iceland, Keldur, Reykjavik, Iceland

Visna, a meningoencephalomyelitis of sheep, was one of the diseases Sigurdsson's concept of slow infections was based on [1]. The causative agent, a retrovirus, was isolated in 1957 [2]. Visna virus is now classified with several animal viruses and the human immunodeficiency virus (HIV) in a subgroup of retroviruses, called lentiviruses, a term quite fittingly derived from Sigurdsson's concept, since visna virus was the first of this group to be isolated.

Visna was eradicated in Iceland several decades ago [3] but we have continued to study the host-virus interactions as both the virus and the disease pose several interesting questions. Furthermore, we are interested in visna, as a model for multiple sclerosis (MS), since visna is also a demyelinating disease, featuring plaques of primary demyelination resembling those characteristic for MS [4].

More recently the focus has shifted to comparative aspects with AIDS as the causative viruses are related and have several properties in common both in vitro and in vivo. Several aspects of the host reaction in HIV infection bear some resemblance to those observed in visna.

[1]We are indebted to Dr. V. V. Joshi for material from lymphoid interstitial pneumonia in AIDS, Dr. H. Budka and Dr. K. Stefánsson for material from AIDS encephalopathy, Ms. Eygló Gísladóttir for technical assistance and Mrs. Margrét Kristinsdóttir for secretarial help.

The Agents

Because of nucleotide sequence homologies, the causative agents of visna and AIDS are classified together with several other non-oncogenic retroviruses in a separate group called lentiviruses. The morphology of both budding viruses and the complete viral particles is practically identical for visna virus and HIV and the general organization of the viral genome is similar [5–7]. A characteristic cytopathic feature of both visna virus and HIV in permissive cells in tissue culture is the formation of multinucleated giant cells [8, 9].

Target Organs

Visna was brought to Iceland with imported sheep in 1933 [3]. The main clinical manifestations observed during the epidemic in Iceland were, however, not from the central nervous system (CNS) but from the lungs due to a chronic interstitial pneumonia called in Icelandic maedi, meaning dyspnea. In some sheep flocks, however, the mortality from visna exceeded that from maedi. The causative viruses, first isolated from the CNS of visna-affected sheep [2] and a few years later from the lungs of maedi-affected sheep [10], were shown to be serologically related and early transmission experiments indicated that visna and maedi were but two different manifestations of infection with the same virus [11], an interesting parallel with HIV infection where there is a growing body of evidence that the nervous system and the lungs may be primary target organs [12, 13].

The evolution of a severe CNS affection seems to be unique for Iceland. In other countries where infection with these ovine lentiviruses is endemic, the pulmonary affection dominates, clinical visna is very rare and is usually an accidental finding at autopsy [14, 15]. The occurrence of clinical visna in a few sheep flocks in a restricted area in Iceland has been a puzzle. The importance of a host factor is indicated by the fact that the Icelandic breed of sheep is clearly more susceptible than other breeds. Thus in experimental infections in American sheep using a virus strain, K1514, that regularly causes progressive encephalitis in Icelandic sheep [16], a tenfold higher dose was needed to induce an encephalitis that was self-limiting [17].

Recent preliminary findings in our laboratory by restriction fragment polymorphism analysis [Valgerdur Andrésdóttir, unpubl. results] on several different maedi and visna strains show that the maedi and visna strains differed by 71% in restriction sites, corresponding to an approximately 12% difference in nucleotide sequence. This is comparable to the greatest variation observed for different HIV-1 isolates [18]. Another interesting parallel to HIV was that the maedi and visna viruses were apparently much more stable in tissue culture than during passage in sheep. It has been found that the in vivo mutation rate is most pronounced in the gene for the envelope glycoprotein [7, 19, 20].

The Role of Antigenic Variation in Virus Persistence and Evolution of Lesions

It has been suggested that the emergence in vivo of mutant virus strains which are not neutralized by the sheep's own sera might be an important mechanism in virus persistence and evolution of lesions in visna [21–23]. A detailed analysis of variation of neutralizing epitopes of isolates from 20 of our long-term infected sheep did not support this hypothesis [24]. By isolating 76 virus strains over the period soon after to more than 7 years after infection and testing them against three different reference sera, a high frequency of variants (16%) was detected but the variants were randomly dispersed over time and among different sheep and only in one case did a variant replace the infecting strain. Furthermore, analysis of 27 isolates from the CNS and CSF of 6 sheep with clinical visna against serial sera from the same sheep did not indicate that mutants are important for evolution of pathological lesions, as none of the isolates typed as variants. Other workers came to a similar conclusion [25].

A more likely explanation for virus persistence and the slow evolution of disease is the so-called provirus theory, i.e. that the virus is carried as a provirus integrated into the cellular DNA and rarely expressed [26, 27], is dealt with more thoroughly by Ashley Haase in his chapter.

Host-Virus Interactions

Spread of Infection

Several aspects of the host-virus interactions in visna and AIDS show interesting parallels. Both infections cause a multiorgan systemic disease and in both the virus spreads within the body by the blood stream. Infection with visna virus leads to viremia shortly after infection, regardless of the route of infection [16]. In our experience the infection in the blood is strictly cell-associated, thus virus could never be isolated from the plasma [16]. The viremia persists in spite of high levels of neutralizing and complement-fixing antibodies. Thus in long-term infections we have been able to isolate virus from the blood during an observation period of 10 years [28], which indicates that replication of virus in circulating leukocytes is restricted and that they thus escape destruction by the immune response. This mechanism of spread of infection has been referred to as the Trojan horse mechanism [29].

It has, however, been difficult to identify infected leukocytes, partly for the reason that in visna only 1 in 10^5 to 10^6 of peripheral leukocytes are infected [16], a frequency similar to that found in HIV infection [30]. In American sheep it has been shown that monocytes and monocyte precursors are infected. Replication of the virus is restricted in the blood monocytes but upon migration into the tissues and differentiation into macrophages this restriction is at least partly abrogated. A puzzling observation is that only a certain subpopulation of tissue macrophages was found to be permissive for the infection [23, 31, 32].

In Icelandic sheep we have recently demonstrated expression of viral proteins in macrophages in brain lesions [33] and succeeded in infecting monocytes from the peripheral blood in vitro [unpubl. results]. The interesting observation is that replication of virus in monocytes is apparently different from that found in the traditional permissive choroid plexus cell cultures. Thus according to our preliminary findings, virus buds into cytoplasmic vacuoles in monocytes where an accumulation of viral particles is found whereas budding from the cell membrane was rarely observed, which is in sharp contrast to the replication in permissive choroid plexus cells where budding is exclusively from the cell membrane and viral particles accumulate extracellularly (fig. 1). In infection with HIV, monocytes and monocyte precursors have been found to be infected and display a similar pattern of replication [34, 35]. This could certainly be important in the spread of infection in these diseases, as another form of a Trojan horse mechanism.

Our earlier studies indicated that not only monocytes but also lymphocytes of peripheral blood are permissive for infection [16, 36]. In an immunohistochemical study of the CNS, lymphocytes and plasma cells were found to express viral proteins [33]. Thus as in HIV both lymphocytes and monocytes of the peripheral blood seem to be target cells of infection with visna virus, and are possibly important for dissemination of the infection within the body.

Pathology

In spite of the fundamental difference in the host reaction to infection with HIV and visna virus, that an immunodeficiency is not a feature of infection with visna virus, there are certain parallels, both in the target organs and some of the features of pathological lesions, especially those observed in the initial stages of HIV infection, are reminiscent of those observed in visna.

It is becoming increasingly evident that the CNS is one of the primary target organs in HIV infection and furthermore a lymphoid interstitial pneumonia, resembling maedi, has been reported in children and adults with AIDS [37–40]. In both there is an interstitial infiltration of mononuclear cells and prominent lymphoid hyperplasia (fig. 2).

In visna there is a follicular hyperplasia of lymph nodes, especially in those draining the main target organs, the brain and the lungs. This hyperplasia is similar to that observed in the generalized lymphadenopathy observed in the initial stages of AIDS [41]. In visna, however, a lymphoid depletion comparable to that observed in the later stages of AIDS is not found.

This difference in the host response makes the comparison of the CNS lesions in AIDS and visna more problematic. Thus it has been difficult to sort out what lesions in the CNS in AIDS are primarily due to the virus because of the frequent opportunistic infections and tumors. Furthermore, in AIDS we are looking at an end-stage of a disease prolonged with various therapeutic measures. Because of the reluctance to take brain biopsies we have a very fragmentary knowledge of the evolution of CNS lesions in AIDS. Nevertheless, some of the early lesions found in AIDS encephalopathy are comparable to those observed in visna.

In AIDS we have found perivascular infiltrates which are of similar composition as those characteristic for visna, i.e. consisting of macrophages, lymphocytes and occasional plasma cells (fig. 3). It has been suggested that such perivascular infiltrates may represent the earliest detectable changes in the CNS in HIV infection [42].

Multinucleated giant cells, which have been considered a hallmark of CNS lesions in AIDS [43, 44], have been detected in visna (fig. 4). But in contrast to AIDS they are apparently very rare in visna and have only been found in experiments with a highly neurovirulent virus [33] selected by serial passage in sheep [45]. This may be a reflection of the very low viral titers in visna-affected brains [16]. A further feature in common is the occurrence of glial nodules (fig. 5).

However, the characteristic finding in AIDS subacute encephalopathy or progressive diffuse leukoencephalopathy of a myelin pallor in the white mat-

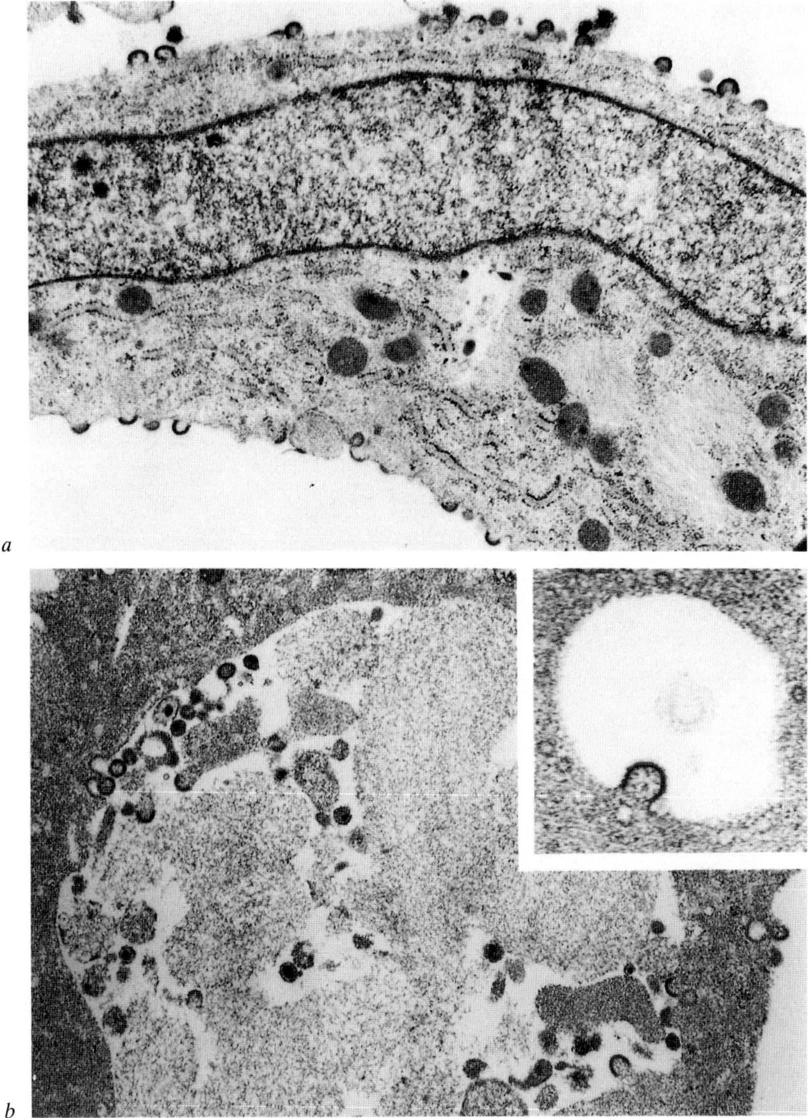

Fig. 1. Tissue culture infected with visna virus. *a* Choroid plexus. Numerous viral particles budding from the plasma membrane. ×17,000; *b* Macrophage. Numerous incomplete and complete viral particles in a cytoplasmic vacuole and several particles budding into the vacuole. ×15,000. Inset: Higher magnification of viral budding into a cytoplasmic vacuole. ×30,000.

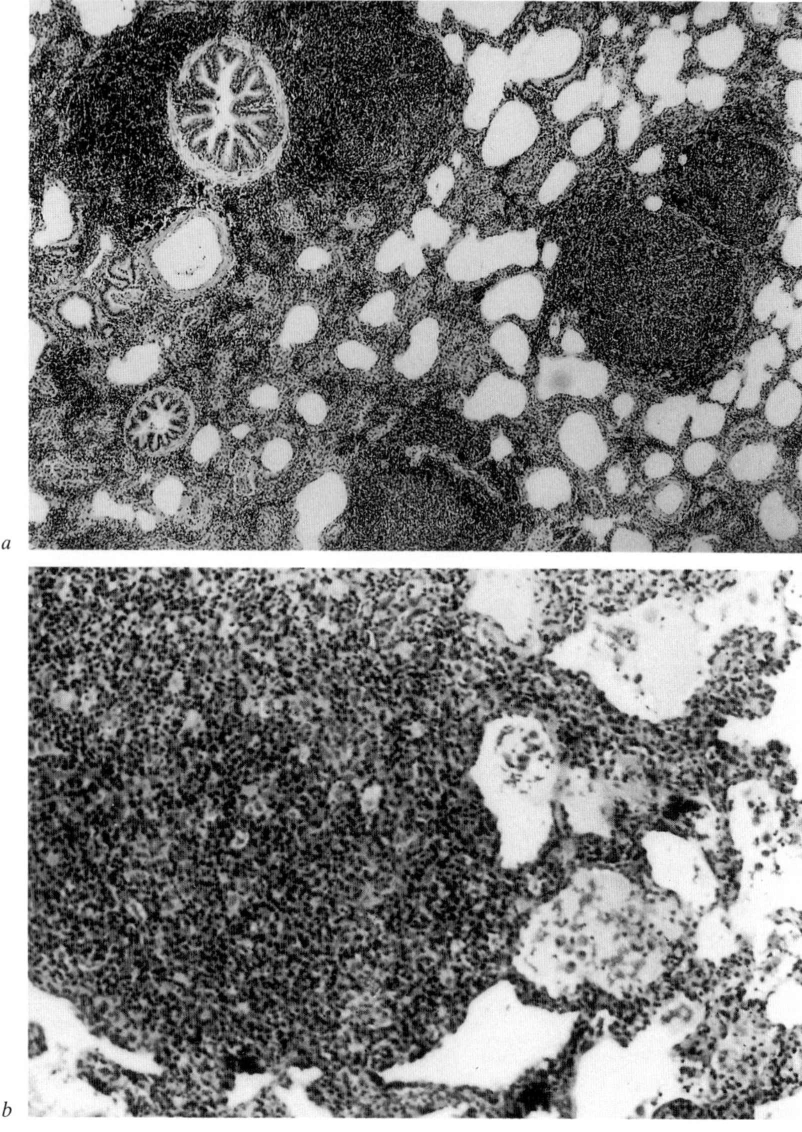

Fig. 2. Lungs. Pronounced lymphoid hyperplasia and infiltration of intraalveolar septa in *(a)* maedi, HE, × 40 and *(b)* AIDS, HE × 95.

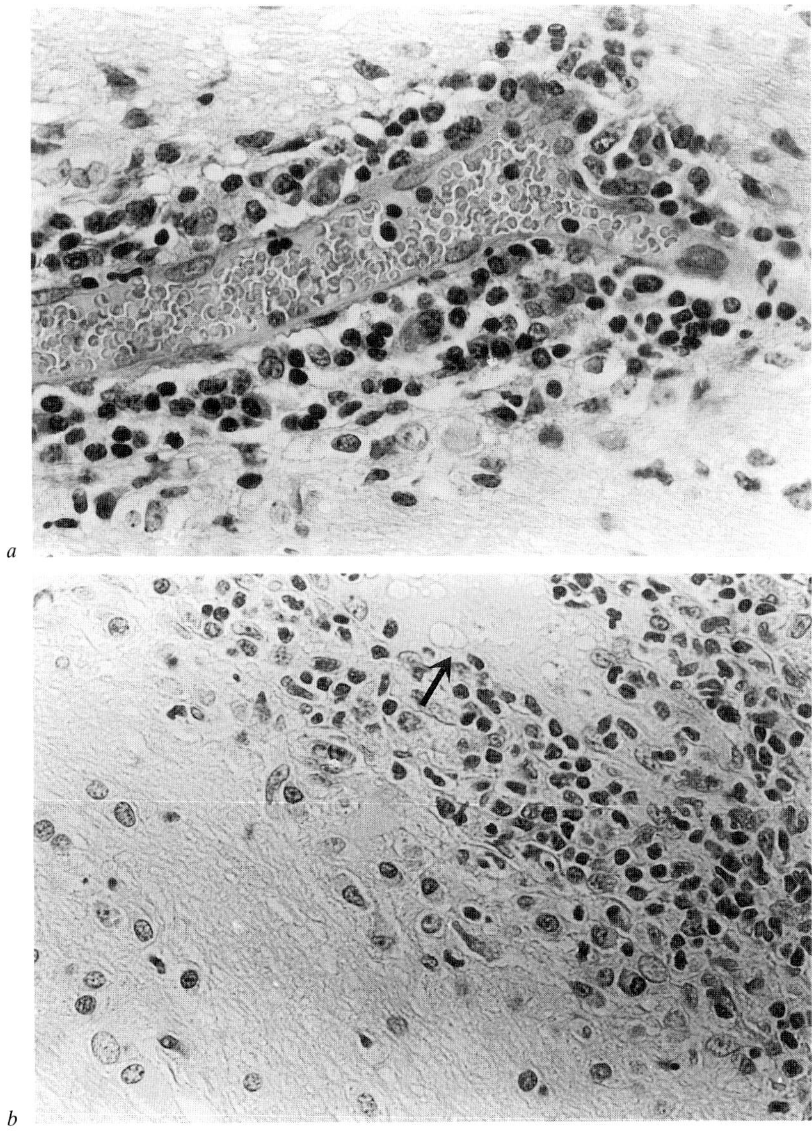

Fig. 3. Brain. Perivascular infiltrates of mononuclear cells in *(a)* visna and *(b)* AIDS. Arrow: Lumen of the vessel. HE. ×380.

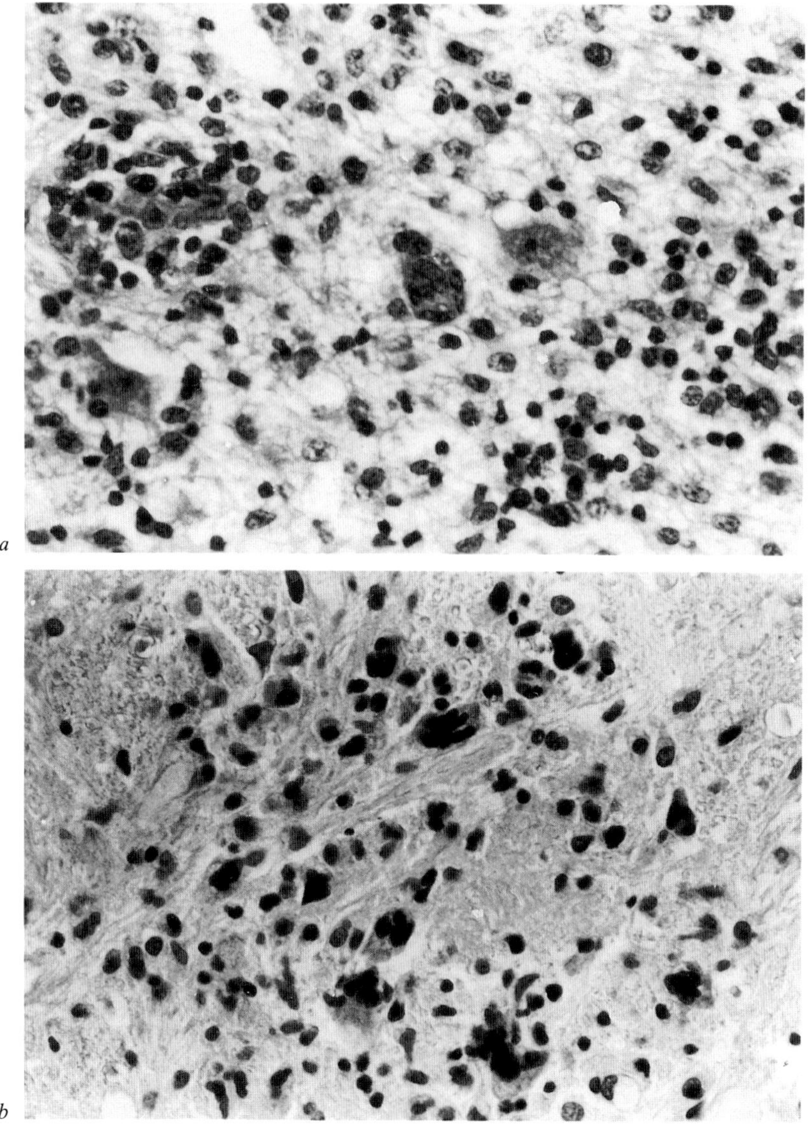

Fig. 4. Brain. Multinucleated giant cells in *(a)* visna and *(b)* AIDS. HE. × 380.

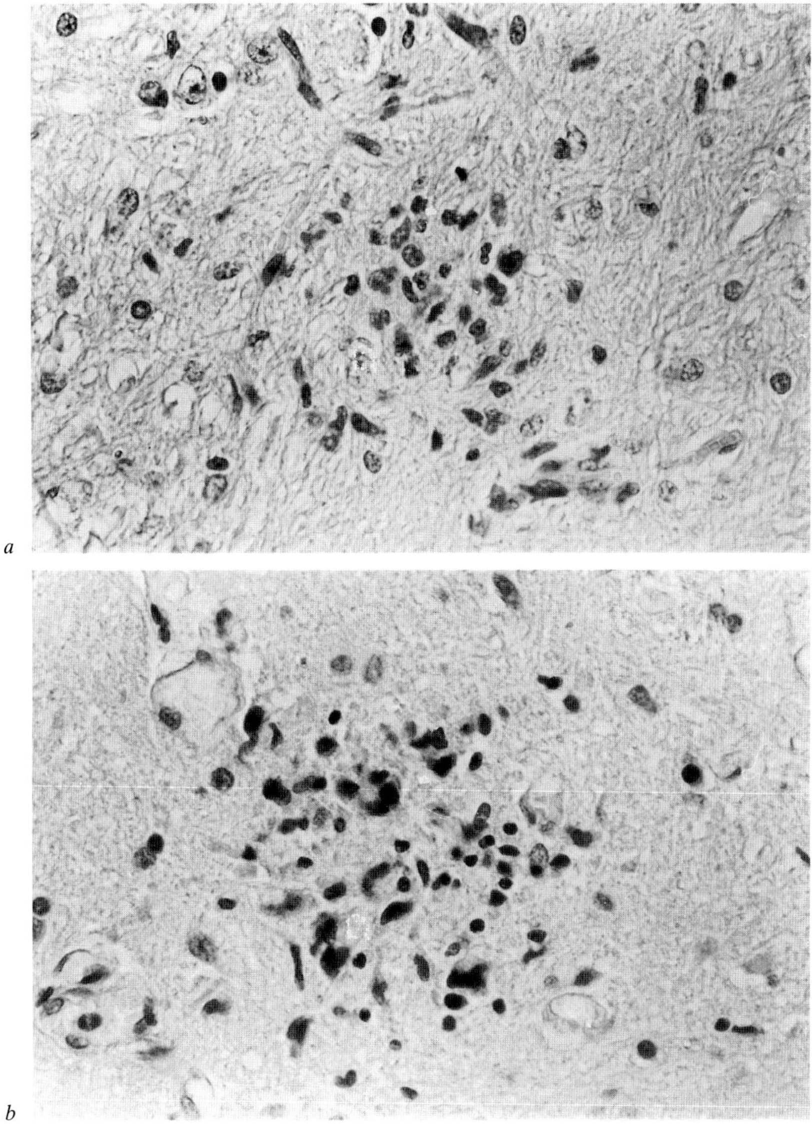

Fig. 5. Brain. Glial nodules in *(a)* visna and *(b)* AIDS. HE. ×380.

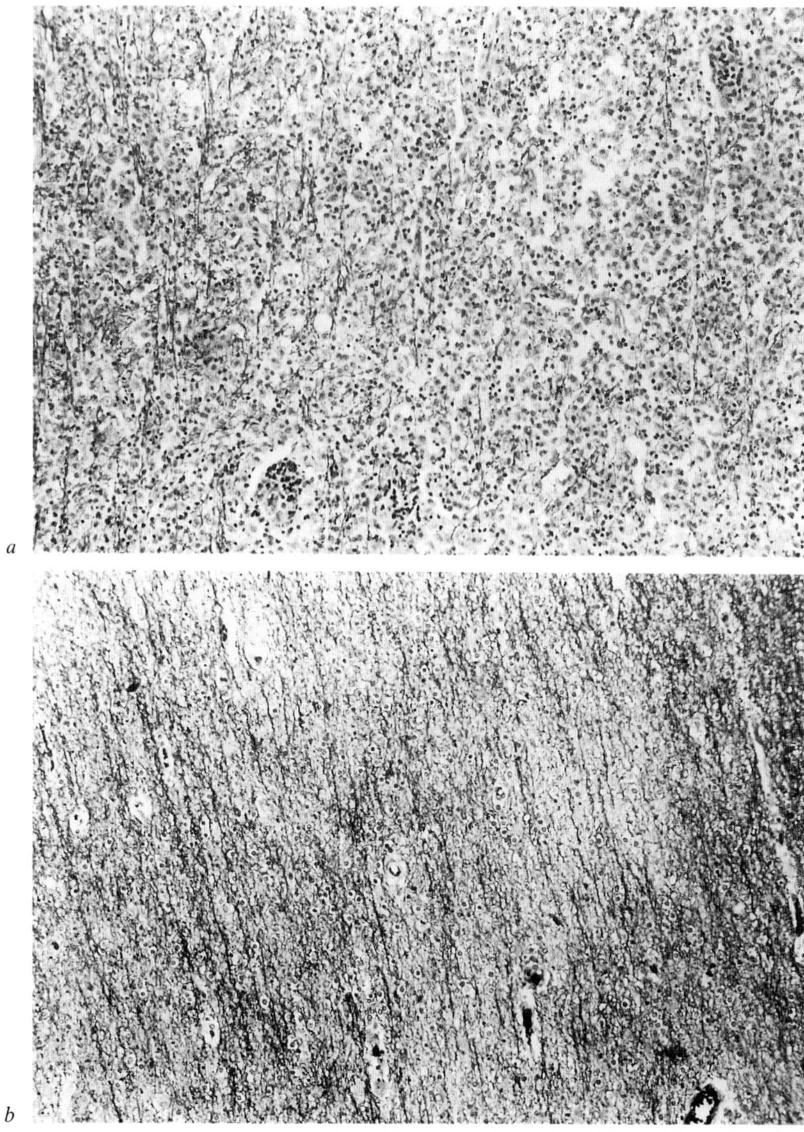

Fig. 6. Brain. Myelin breakdown in *(a)* visna, with dense infiltration of inflammatory cells and *(b)* AIDS. Klüver-Barrera. ×95.

ter due to rarification of the myelin, and sparse myelophages but no inflammatory reaction, is quite different from the myelin breakdown usually seen in visna, where there is a diffuse infiltration of inflammatory cells (fig. 6). A further difference is that in visna a pronounced lymphoid proliferation is frequently observed especially in the choroid plexus. This change as well as the similar lesions in the lungs and hyperplasia of lymph nodes led to the suggestion that the lesions in visna might be immunopathological.

Pathogenesis and Target Cells in the CNS

By applying immunosuppression [46] and immunopotentiation [47] we were able to demonstrate that at least the early lesions were immune-mediated. Furthermore, our findings indicated that the immune response was not directed against myelin antigens [48] but virus-induced antigens [16, 36].

It proved, however, difficult to demonstrate what cells were the targets of the damaging immune response, i.e. what cells in the CNS were permissive for infection. The reason was that apparently very few cells are infected and virus replication is severely restricted [26, 27]. In an attempt to increase the possibility of detecting infected cells in the CNS we selected for a more neurovirulent strain by serial passage through sheep and came up with a strain, K1772, that caused clinical disease in approximately 50% of sheep [45] within 6 months which is in sharp contrast to the prolonged incubation period following infection with the standard K1514 strain [28].

Both in situ hybridization for detection of viral RNA and immunohistochemistry for viral proteins were done on brain sections of sheep infected with this strain. By a combined immunohistochemical staining with cell markers and in situ hybridization for viral RNA, infection of oligodendrocytes and astrocytes was detected [49], however, most of the cells harboring the viral genome could not be identified.

An extensive study of the expression of viral antigens was done on brains from four cases infected with neurovirulent virus strains, applying mouse monoclonal antibodies against two different core proteins of the virus [33]. Antigen-positive cells were rare in the neuroparenchyme but more numerous in the choroid plexus, where approximately 1 in 1,000 cells expressed viral proteins. A wide variety of cells were found to express these core proteins of the visna virus. Inflammatory cells, macrophages, lymphocytes and an occasional plasma cell were the most common antigen-positive cells, but in addition viral proteins were detected in endothelial cells, fibroblasts, pericytes, choroidal epithelial cells and an occasional meningeal cell. Viral antigen could, however, not be detected in glial cells. This difference from the results of the in situ hybridization may be due to different sensitivities of the methods or an indication of a posttranscriptional block in infected astrocytes and oligodendrocytes, a view supported by the fact that there is a severe restriction of virus replication in the infected host [26, 27].

The spectrum of infected cells in the CNS of visna-infected sheep detected by both in situ hybridization for the viral genome and immunohistochemistry for viral proteins resembles that reported in the CNS in AIDS by applying these methods [50–53] with the exception that infection of choroidal epithelial cells has not been described in HIV infection and we have not detected infection in microglial cells, which according to recent reports may be the main target cell expressing viral proteins in AIDS [54, 55]. However, more studies are needed in visna for example with double immunostaining with cell markers and for viral proteins to clarify the possible role of microglial cells. Interestingly, we were able to stain a few cells, mainly inflammatory, by applying our monoclonal antibodies against core protein of visna virus on sections from the brain of an AIDS patient [56].

Conclusion

It is clear that in addition to the relationship of visna virus and HIV and similar characteristics of their behavior in vitro, the host-virus interactions in infections with these viruses show several interesting parallels, such as target cells in the blood, mechanism of virus spread within the body and to some extent the target organs of infection. Although immunodeficiency is not a feature of visna virus infection, the pathological changes show some resemblance especially to those found in early stages of infection with HIV. The spectrum of target cells in the CNS in both infections is apparently similar. As there is increasing evidence that the nervous system is one of the primary target organs of infection with HIV, visna of sheep offers a promising possibility as a model for trials aimed at preventing or treating AIDS.

References

1 Sigurdsson, B.: Observations on three slow infections of sheep. Br. Vet. J. *110:* 255–270 (1954).
2 Sigurdsson, B.; Thormar, H.; Pálsson, P. A.: Cultivation of visna virus in tissue culture. Arch. Ges. Virusforsch. *10:* 368–381 (1960).
3 Pálsson, P. A.: Maedi and visna in sheep; in Kimberlin, R. H. (ed.): Slow Virus Diseases of Animals and Man, pp. 17–43 (North-Holland, Amsterdam 1976).

4 Georgsson, G.; Martin, J. R.; Klein, J.; Pálsson, P. A.; Nathanson, N.; Pétursson, G.: Primary demyelination in visna: An ultrastructural study of Icelandic sheep with clinical signs following experimental infection. Acta Neuropathol. (Berl.) *57:* 171–178 (1982).

5 Gonda, M. A.; Wong-Staal, F.; Gallo, R. C.; Clements, J. E.; Narayan, O.; Gilden, R. V.: Sequence homology and morphologic similarity of HTLV-III and visna virus, a pathogenic lentivirus. Science *227:* 170–173 (1985).

6 Sonigo, P.; Alizon, M.; Staskus, K.; Klatzmann, D.; Cole, S.; Danos, O.; Retzel, E.; Tiollais, P.; Haase, A.; Wain-Hobson, S.: Nucleotide sequence of the visna lentivirus: Relationship to the AIDS virus. Cell *42:* 369–382 (1985).

7 Braun, M. J.; Clements, J. E.; Gonda, M. A.: The visna virus genome: Evidence for a hypervariable site in the *env* gene and sequence homology among lentivirus envelope proteins. J. Virol. *61:* 4046–4054 (1987).

8 Thormar, H.: Visna-maedi infection in cell cultures and in laboratory animals; in Kimberlin, R. H. (ed.): Slow Virus Diseases of Animals and Man, pp. 97–114 (North-Holland, Amsterdam 1976).

9 Popovic, M.; Sarngadharan, M. G.; Read, E.; Gallo, R. C.: Detection, isolation and continuous production of cytopathic retroviruses (HTLV-III) from patients with AIDS and pre-AIDS. Science *224:* 497–500 (1984).

10 Sigurdardóttir, B.; Thormar, H.: Isolation of a viral agent from the lungs of sheep affected with maedi. J. Infect. Dis. *114:* 55–60 (1964).

11 Gudnadóttir, M.; Pálsson, P. A.: Successful transmission of visna by intrapulmonary inoculation. J. Infect. Dis. *115:* 217–225 (1965).

12 Price, R. W.; Brew, B.; Sidtis, J.; Rosenblum, M.; Scheck, A. C.; Cleary, P.: Central nervous system HIV-1 infection and AIDS dementia complex. Science *239:* 586–592 (1988).

13 Chayt, K.; Harper, M. E.; Marselle, L. M.; Lewin, E. B.; Rose, R. M.; Oleske, J. M.; Epstein, L. G.; Wong-Staal, F.; Gallo, R. C.: Detection of HTLV-III RNA in lungs of patients with AIDS and pulmonary involvement. JAMA *256:* 2356–2359 (1986).

14 De Boer, G. F.; Terpstra, C.; Houwers, D. J.: Studies in epidemiology of maedi/visna in sheep. Res. Vet. Sci. *26:* 202–208 (1979).

15 Cutlip, R. C.; Jackson, T. A.; Lemkuhl, H. D.: Lesions of ovine progressive pneumonia: Interstitial pneumonitis and encephalitis. Am. J. Vet. Res. *40:* 1370–1374 (1979).

16 Pétursson, G.; Nathanson, N.; Georgsson, G.; Panitch, H.; Pálsson, P. A.: Pathogenesis of visna. I. Sequential virologic, serologic and pathologic studies. Lab. Invest. *35:* 402–412 (1976).

17 Narayan, O.; Strandberg, J. D.; Griffin, D. E.; Clements, J. E.; Adams, R. J.: Aspects of the pathogenesis of visna in sheep; in Mims, C. A., Cuzner, M. L., Kelly, R. E. (eds): Viruses and Demyelinating Diseases, pp. 125–140 (Academic Press, London 1983).

18 Saag, M. S.;Hahn, B. H.; Gibbons, J.; Li, Y.; Parks, E. S.; Parks, W. P.; Shaw, G. M.: Extensive variation of human immunodeficiency virus type-1 in vivo. Nature *334:* 440–444 (1988).

19 Clements, J. E.; Pedersen, F. S.; Narayan, O.: Genomic changes associated with antigenic variation of visna virus during persistent infection. Proc. Natl. Acad. Sci. USA *77:* 4454–4458 (1980).

20 Cheng-Mayer, C.; Homsy, J.; Evans, L. A.; Levy, J. A.: Identification of human immunodeficiency virus subtypes with distinct patterns of sensitivity to serum neutralization. Proc. Natl. Acad. Sci. USA *85:* 2815–2819 (1988).

21 Gudnadóttir, M.: Visna-maedi in sheep. Prog. Med. Virol. *18:* 336–349 (1974).
22 Narayan, O.; Griffin, D.E.; Chase, J.: Antigenic shift of visna virus in persistently infected sheep. Science *197:* 376–378 (1977).
23 Narayan, O.; Wolinsky, J.S.; Clements, J.E.; Strandberg, J.D.; Griffin, D.E.; Cork, L.C.: Slow virus replication: The role of macrophages in the persistence and expression of visna viruses of sheep and goats. J. Gen. Virol. *59:* 345–356 (1982).
24 Lutley, R.; Pétursson, G.; Pálsson, P.A.; Georgsson, G.; Klein, J.; Nathanson, N.: Antigenic drift in visna: Virus variation during long-term infection of Icelandic sheep. J. Gen. Virol. *6:* 1433–1440 (1983).
25 Thormar, H.; Barshatzky, M.R.; Arnesen, K.: Kozlowski, P.B.: The emergence of antigenic variants is a rare event in long-term visna virus infection in vivo. J. Gen. Virol. *64:* 1427–1432 (1983).
26 Haase, A.T.: The AIDS lentivirus connection. Microb. Pathog. *1:* 1–4 (1986).
27 Haase, A.T.: Pathogenesis of lentivirus infections. Nature *322:* 130–136 (1986).
28 Georgsson, G.; Pálsson, P.A.; Pétursson, G.: Pathogenesis of visna; in Serlupi Crescenzi, G. (ed).: A Multidisciplinary Approach to Myelin Diseases, pp. 303–318 (Plenum Press, New York 1987).
29 Peluso, R.; Haase, A.; Stowring, L.; Edwards, M.; Ventura, P.: A Trojan horse mechanism for the spread of visna in monocytes. Virology *147:* 2231–2236 (1985).
30 Harper, M.E.; Marselle, L.M.; Gallo, R.C.; Wong-Staal, F.: Detection of lymphocytes expressing human T-lymphotropic virus type III in lymph nodes and peripheral blood from infected individuals by in situ hybridization. Proc. Natl. Acad. Sci. USA *83:* 772–776 (1986).
31 Gendelman, H.E.; Narayan, O.; Molineux, S.; Clements, J.E.; Ghotbi, Z.: Slow persistent replication of lentiviruses: Role of tissue macrophages and macrophage precursors in bone marrow. Proc. Natl. Acad. Sci. USA *82:* 7086–7090 (1985).
32 Gendelman, H.E.; Narayan, O.; Kennedy-Stoskopf, S.; Kennedy, P.G.E.; Ghotbi, Z.; Clements, J.E.; Stanley, J.; Pezeshkpour, G.H.: Tropism of sheep lentivirus for monocytes. J. Virol. *58:* 67–74 (1986).
33 Georgsson, G.; Houwers, D.J.; Pálsson, P.A.; Pétursson, G.: Expression of viral antigens in the central nervous system of visna-infected sheep: an immunohistochemical study on experimental visna induced by virus strains of increased neurovirulence. Acta Neuropathol. (Berl.) *77:* 299–306 (1989).
34 Gartner, S.; Markovits, P.; Markovitz, D.M.; Kaplan, M.H.; Gallo, R.C.; Popovic, M.: The role of mononuclear phagocytes in HTLV-III/LAV infection. Science *233:* 215–219 (1986).
35 Folks, T.M.; Kessler, S.W.; Orenstein, J.M.; Justement, J.S.; Jaffe, E.S.; Fauci, A.S.: Infection and replication of HIV-1 in purified progenitor cells of normal human bone marrow. Science *247:* 919–922 (1988).
36 Pétursson, G.; Martin, J.R.; Georgsson, G.; Nathanson, N.; Pálsson, P.A.: Visna, the biology of the agent and the disease; in Tyrrell, D.A.J. (ed.): New Perspectives in Clinical Microbiology. Aspects of Slow and Persistent Virus Infections, pp. 198–220 (Nijhoff, The Hague 1979).
37 Joshi, V.V.; Oleske, J.M.; Minnefor, A.B.; Saad, S.; Klein, K.M.; Sing, R.; Zabala, M.; Dadzie, C.; Simpser, M.; Rapkin, R.H.: Pathologic pulmonary findings in children with the acquired immunodeficiency syndrome. A study of ten cases. Hum. Pathol. *16:* 241–246 (1985).

38 Marchevsky, A.; Rosen, M. J.; Chrystal, G.; Kleinerman, J.: Pulmonary complication of the acquired immunodeficiency syndrome. A clinicopathologic study of 70 cases. Hum. Pathol. *16:* 659–670 (1985).
39 Grieco, M. H.; Chinoy-Acharya, P.: Lymphocytic interstitial pneumonia associated with the acquired immune deficiency syndrome. Am. Rev. Respir. Dis. *131:* 952–955 (1985).
40 Oleske, J.; Minnefor, A.; Cooper, R., Jr.: Thomas, K.; dela Cruz, A.; Ahdieh, H.; Guerrero, I.; Joshi, V. V.; Desposito, F.: Immune deficiency syndrome in children. JAMA *249:* 2345–2349 (1983).
41 Rácz, P.; Tenner-Rácz, K.; Kahl, C.; Feller, A. C.; Kern, P.; Dietrich, M.: Spectrum of morphologic changes of lymph nodes from patients with AIDS or AIDS-related complexes. Prog. Allergy *37:* 81–181 (1986).
42 Lantos, P. L.; McLaughlin, J. E.; Scholtz, C. L.; Berry, C. L.; Tighe, J. R.: Neuropathology of the brain in HIV infection. Lancet *i:* 309–311 (1989).
43 Budka, H.: Multinucleated giant cells in brain: a hallmark of acquired immune deficiency syndrome (AIDS). Acta Neuropathol. (Berl.) *69:* 253–258 (1986).
44 Sharer, L. R.; Cho, E.-S.; Epstein, L. G.: Multinucleated giant cells and HTLV-III in AIDS encephalopathy. Hum. Pathol. *16:* 760 (1985).
45 Lutley, R.; Pétursson, G.; Georgsson, G.; Pálsson, P. A.; Nathanson, N.: Strains of visna virus with increased neurovirulence; in Sharp, J. M., Jørgensen, R. H. (eds): Slow Viruses in Sheep, Goats and Cattle, pp. 45–49 (Commission of the European Communities, Luxembourg 1985).
46 Nathanson, N.; Panitch, H.; Pálsson, P. A.; Pétursson, G.; Georgsson, G.: Pathogenesis of visna. II. Effect of immunosuppression upon early central nervous system lesions. Lab. Invest. *35:* 444–451 (1976).
47 Nathanson, N.; Martin, J. R.; Georgsson, G.; Pálsson, P. A.; Lutley, R. E.; Pétursson, G.: The effect of post-infection immunization on the severity of experimental visna. J. Comp. Pathol. *91:* 185–191 (1981).
48 Panitch, H.; Pétursson, G.; Georgsson, G.; Pálsson, P. A.; Nathanson, N.: Pathogenesis of visna. III. Immune response to central nervous system antigens in experimental allergic encephalomyelitis and visna. Lab. Invest. *35:* 452–460 (1976).
49 Stowring, L.; Haase, A. T.; Pétursson, G.; Georgsson, G.; Pálsson, P. A.; Lutley, R.; Roos, R.; Szuchet, S.: Detection of visna virus antigens and RNA in glial cells in foci of demyelination. Virology *141:* 311–318 (1985).
50 Pumarola-Sune, T.; Navia, B. A.; Cordon-Cardo, C.; Cho, E.-S.; Price, R. W.: HIV antigen in the brains of patients with AIDS dementia complex. Ann. Neurol. *21:* 490–496 (1987).
51 Stoler, M. H.; Eskin, T. A.; Benn, S.; Angerer, R. C.; Angerer, L. M.: Human T-cell lymphotropic virus type III infection of the central nervous system. A preliminary in situ analysis. JAMA *251:* 2360–2364 (1986).
52 Wiley, C. A.; Schrier, R. D.; Nelson, J. A.; Lampert, P. W.; Oldstone, M. B. A.: Cellular localization of human immunodeficiency virus infection within the brains of acquired immune deficiency syndrome patients. Proc. Natl. Acad. Sci. USA *83:* 7089–7093 (1986).
53 Koenig, S.; Gendelman, H. E.; Orenstein, J. M.; Dal Canto, M. C.; Pezeshkpour, G. H.; Yungbluth, M.; Janotta, F.; Aksamit, A.; Martin, M. A.; Fauci, A. S.: Detection of AIDS virus in macrophages in brain tissue from AIDS patients with encephalopathy. Science *233:* 1089–1093 (1986).

54 Vazeux, R.; Brousse, N.; Jarry, A.; Henin, D.; Marche, C.; Vedrenne, C.; Mikol, J.; Wolff, M.; Michon, C.; Rozenbaum, W.; Bureau, J.-F.; Montagnier, C.; Brahic, M.: AIDS subacute encephalitis: identification of HIV-infected cells. Am. J. Pathol. *126:* 403–410 (1987).
55 Michaels, J.; Price, R. W.; Rosenblum, M. K.: Microglia in the giant cell encephalitis of acquired immune deficiency syndrome. Proliferation, infection and fusion. Acta Neuropathol. (Berl.) *76:* 373–379 (1988).
56 Georgsson, G.; Houwers, D. J.; Stefánsson, K.; Pálsson, P. A.; Pétursson, G.: Immunohistochemical staining of cells in the brain of a patient with a monoclonal antibody to visna virus. Acta Neuropathol. (Berl.) *73:* 406–408 (1987).

Gudmundur Georgsson, MD, Institute for Experimental Pathology,
University of Iceland, Keldur, PO Box 8540, IS–128 Reykjavik (Iceland)

Racz P, Haase AT, Gluckman JC (eds): Modern Pathology of AIDS and Other Retroviral Infections. Basel, Karger, 1990, pp 99–109

HIV and EBV Expression in Lymph Nodes: Immunohistology and in situ Hybridization[1]

Carlo D. Baroni, Stefania Uccini[2]

Second Chair of Pathological Anatomy, Department of Biopathology, University 'La Sapienza', Rome, Italy

Epstein-Barr virus (EBV) is associated with the development of B cell lymphoproliferative diseases (BCLD) in immunocompromised patients. B lymphocytes are infected and immortalized both in vivo [1] and in vitro [2] by EBV, thus resulting in polyclonal proliferation and activation. It is now well known that in normal subjects EBV-induced proliferation and activation of B lymphocytes is mainly controlled by T cells [3, 4], by natural killer cell activity [5] and by antibody-dependent cellular cytoxicity [6]. On the contrary, in patients affected by immunodeficiency, the EBV-driven B lymphocyte proliferation can go uncontrolled. This, in turn, may result in the evolution from a polyclonal to a monoclonal B cell proliferation and ultimately in a B cell lymphoma [7]. B cell malignancies have been described in primary immunodeficiencies [8–10] in organ transplant recipients [11, 12] and in patients affected by AIDS [13, 14]. It is well known that AIDS is frequently preceded by persistent generalized lymphadenopathy (PGL) exhibiting the histologic features of a follicular hyperplasia with a peculiar activation and proliferation of B cells populating germinal centers. A number of patients affected by PGL develop BCLD, therefore PGL can perhaps be interpreted as a predisposing condition resulting in the development of BCLD. However, the relationships between PGL and BCLD are still intriguing. The role of the human immunodeficiency virus (HIV) in the etiopathogenesis of AIDS and related syndromes is now well established [15, 16]; on the contrary, HIV-specific sequenc-

[1]Supported by AIRC (Associazione Italiana Ricerca sul Cancro) and by Ministero della Sanità, Progetto AIDS 87.003.

[2]The authors thank Prof. G. R. Pearson and Dr. R. C. Gallo for providing the anti-EBV and HIV monoclonal antibodies, Mr. F. Marziali for excellent technical assistance and Miss M. G. Saladino for manuscript preparation.

es were never detected in AIDS-related BCLD thus indicating that HIV is perhaps not directly involved in B cell lymphomagenesis although it seems to play some role in the pathogenesis of PGL [17–20]. The recent demonstration that EBV is integrated in the DNA of Burkitt's leukemia cells affecting two HIV-positive hemophilic brothers [21] adds further evidence that EBV is closely associated with the development of BCLD in HIV-positive patients. To assess whether or not EBV may also play a role in the pathogenesis of PGL, we have analyzed the localization and frequencies of cells infected by EBV in the lymph nodes of a large group of HIV-positive patients affected by PGL.

Materials and Methods

Lymph node tissue samples selected for the present study included specimens from 50 HIV-positive patients affected by PGL in the hyperplastic phase (32 males and 18 females from 2 to 44 years of age). PGL with hyperplastic changes is histologically classified as: (a) follicular hyperplasia (FH) characterized by increased number of irregularly shaped and activated germinal centers, and (b) follicular hyperplasia with follicular fragmentation (FH+FF) characterized by hemorrhages, lymphocyte infiltration and follicular dendritic reticulum cell lysis of germinal centers.

The following monoclonal antibodies were used to stain acetone-fixed cryostat sections of lymph nodes: DRC-1 (Dakopatts, Denmark) against follicular dendritic reticulum cells present in germinal centers, L2 and R63.2 against EBV-viral capsid antigen (VCA) [22], EBV-early antigen (EA) [23], and M26 directed against HIV p24 core protein [24]. The immune reaction product was revealed using an avidin-biotin peroxidase complex. EBV genome was investigated by in situ hybridization on cryostat sections of 30 lymph nodes, which were hybridized with a biotinylated EBV-DNA probe (Bioprobe®, Enzo Biochem., N.Y., USA); the reaction was enzymatically detected with an avidin-biotin peroxidase complex. Furthermore, EBV genome was studied by Southern blot hybridization of DNA extracted from seven frozen lymph nodes with an EBV cloned probe (EBV Bam HI-W fragment, about 3.07 kb long), nick translated with ^{32}P-labelled deoxycytidine-5'-triphosphate. The following controls were used: (1) HIV-infected H9 cell line; (2) centrifuge pellets of EBV producer B95-8 cell line, P3HR-1-infected Raji cells and EBV nonproducer Raji cells.

Cell suspension studies were performed on cytocentrifuge smears obtained from peripheral blood (PBL). The smears were immunostained and in situ hybridized as previously described for cryostat sections.

Results

Lymph nodes obtained from HIV-positive patients affected by PGL with hyperplastic changes and with or without follicular fragmentation (fig. 1a, b) were immunostained with DRC-1, anti-EBV-VCA, anti-EBV-EA and anti-

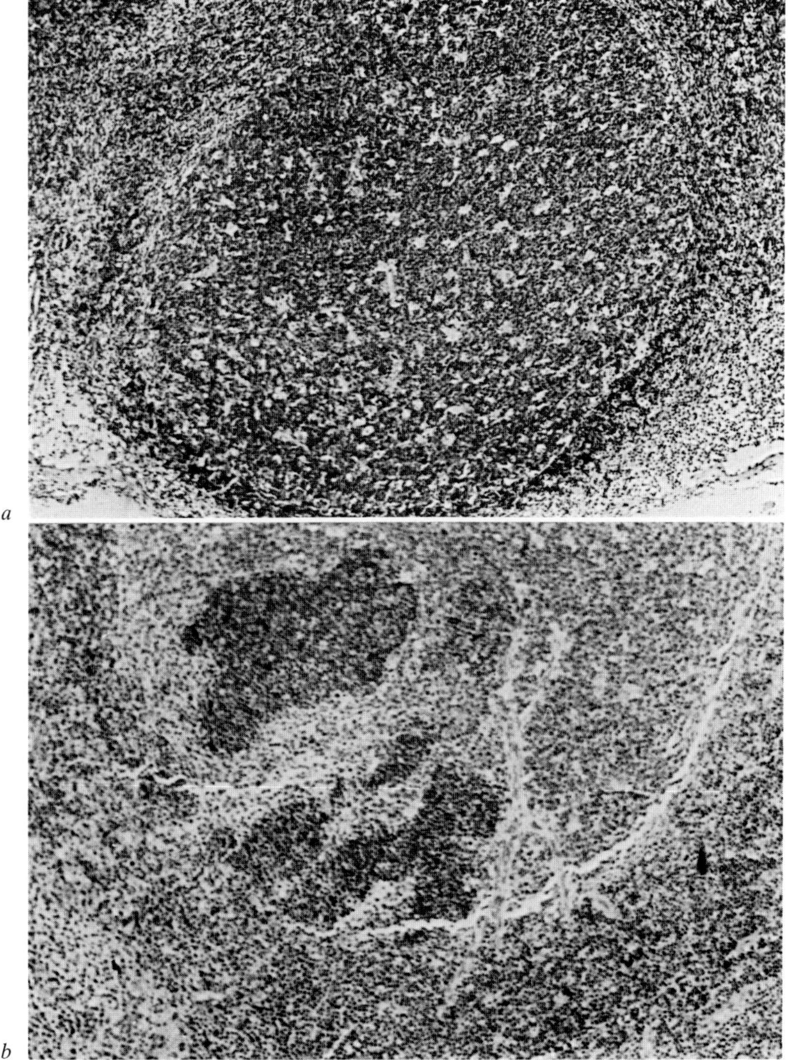

Fig. 1. Lymph node paraffin sections from patients affected by PGL. *a* Follicular hyperplasia with prominent activation of a germinal center and starry-sky appearance. The mantle zone is partly absent. HE. ×120. *b* Follicular hyperplasia with follicular fragmentation indicated by the infiltration of a germinal center by darkly stained small lymphocytes. HE. ×120.

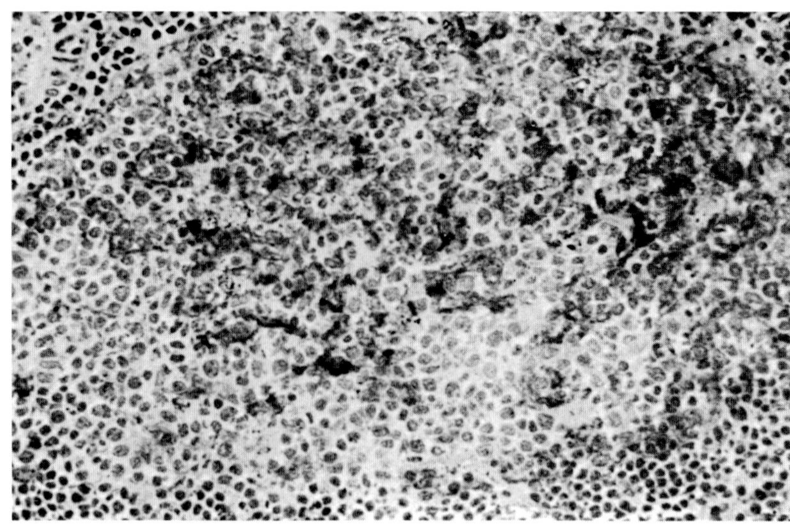

Fig. 2. Lymph node cryostat section immunostained with an anti-p24 HIV core protein monoclonal antibody. The positivity is confined to the germinal centers with a reticular pattern mimicking that of follicular dendritic reticulum cells. ×240.

Table 1. Serology, histology, immunohistochemistry and in situ hybridization of lymph nodes obtained from HIV-positive patients affected by PGL

Patients	Serology		Histology		LN[1] antigens		LN genome	
	HIV	EBV	FH	FH + FF	HIV	EBV	HIV	EBV
50 (32M + 18F)	50/50[2]	42/50	26/50	24/50	43/50	2/50	+/–[1]	0/30

[1] Only few cells in the lymph nodes (LN) are HIV genome-positive.
[2] Number of positive patients/total number of patients.

HIV p24 core protein monoclonal antibodies (table 1). Immunopositivity for HIV p24 antigen was observed in germinal centers of 43 out of 50 lymph nodes having a reticular pattern (fig. 2)., mimicking that of follicular dendritic reticulum cells (fig. 3a, b). In only 2 out of 50 lymph nodes, a few centrocyte-like cells present in germinal centers demonstrated an evident positivity for EBV antigens (fig. 4). The presence of EBV genome was investigated by in situ

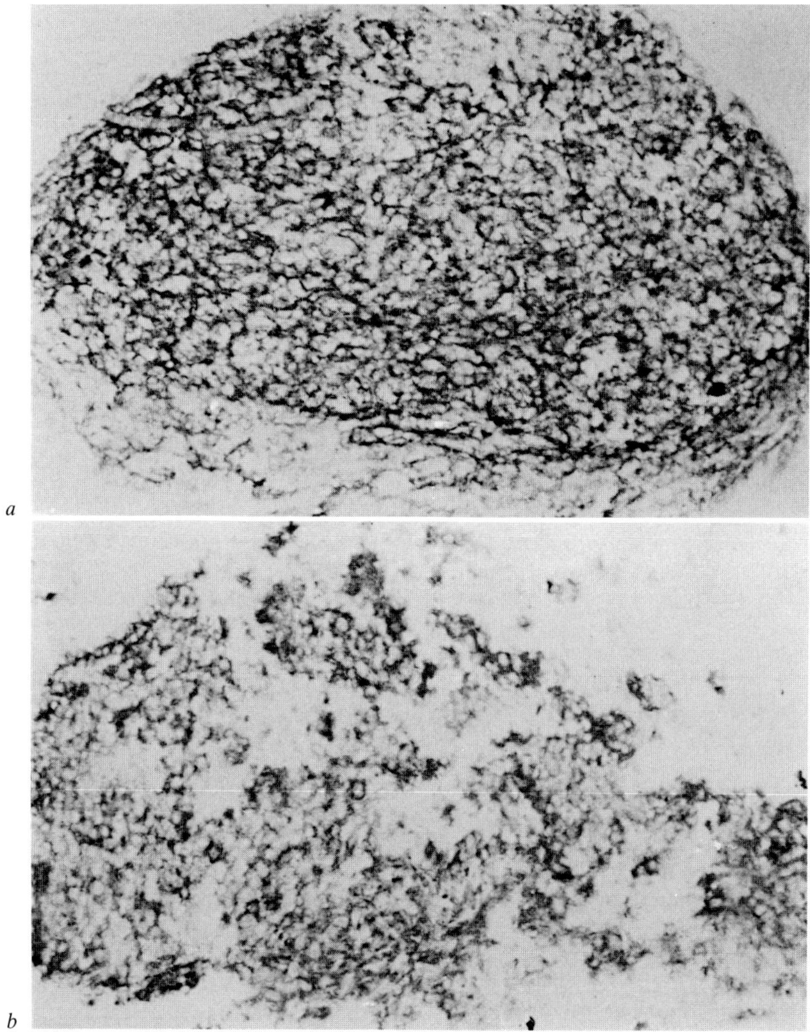

Fig. 3. Lymph node cryostat sections immunostained with DRC-1 MoAb. *a* The processes of the follicular dendritic reticulum cells are markedly DRC-1-positive and appear as a reticular meshwork occupying the entire germinal center. ×240. *b* The processes of the follicular dendritic reticulum cells are disrupted as a result of focal follicular lysis. ×240.

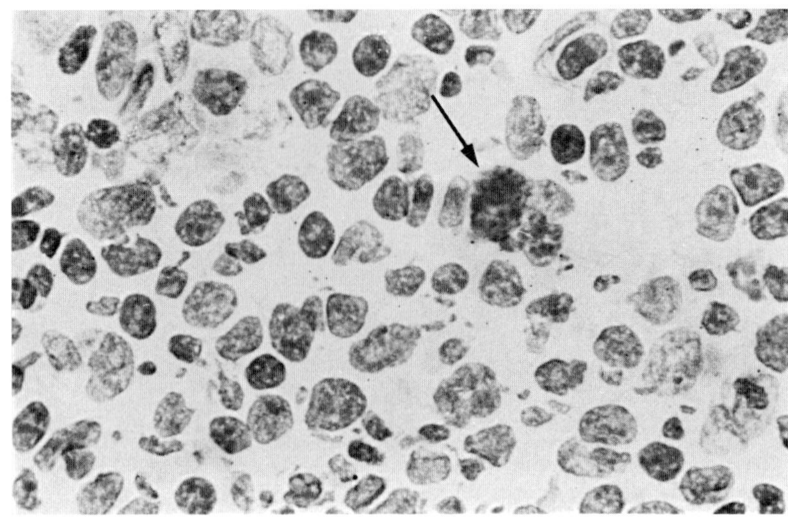

Fig. 4. Lymph node cryostat section immunostained with an anti-EBV-VCA MoAb. An EBV-VCA-positive cell (arrow) is present in a germinal center. × 970.

Table 2. Controls for immunohistochemistry and in situ hybridization

Controls	Antigens			In situ	
	HIV	EBV-EA	EBV-VCA	HIV	EBV
HIV-infected H9	45[1]	–	–	+	–
EBV producer B95-8	–	20–30	20–30	–	15–30
P3HR-1-infected Raji	–	5	–	–	+
EBV nonproducer Raji	–	–	–	–	+
K562	–	–	–	–	–

[1] Percentage of positive cells.

hybridization and by Southern blot analysis; interestingly enough, the EBV genome could never be demonstrated at tissue level.

Cell lines, used as positive controls, are listed in table 2. Between 20 and 30% of the B95-8 cells were immunostained with anti-EBV-VCA and EBV-EA monoclonal antibodies (fig. 5). Furthermore, 15–30% of the B95-8 cells showed a nuclear positivity after EBV-DNA in situ hybridization (table 1, fig. 6). Approximately 5% of P3HR-1-infected Raji cells expressed EBV-EA;

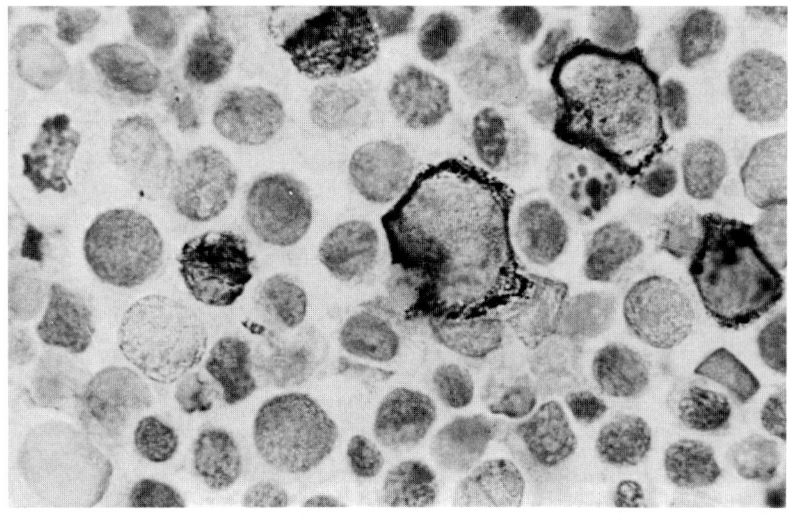

5

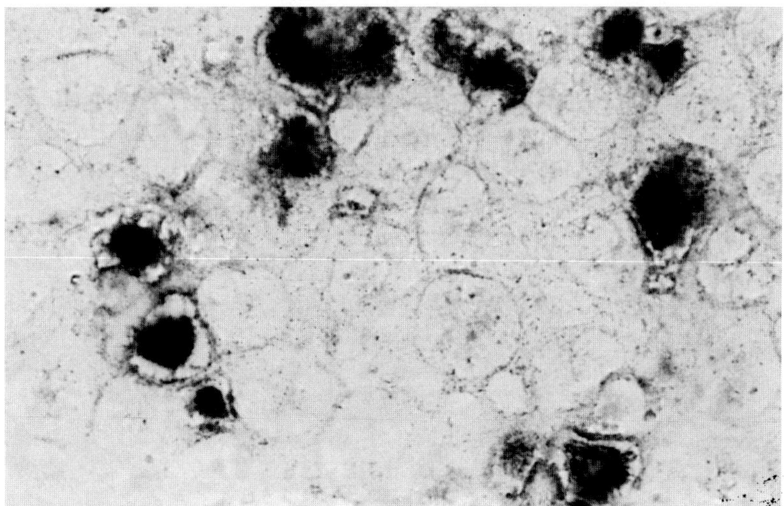

6

Fig. 5. A cytocentrifuge smear prepared from the EBV-infected B95-8 cell line, immunostained with an anti-EBV-VCA MoAb. Numerous cells show a marked positivity appearing as an evident cytoplasmic rim. ×970.

Fig. 6. Cryostat section of a centrifuge pellet prepared from EBV-infected B95-8 cells, in situ hybridized with a biotinylated EBV-DNA probe, enzymatically detected. Numerous cells show an evident nuclear positivity. ×970.

both noninfected Raji cells and P3HR-1-infected Raji cells were positive for in situ hybridization with EBV-DNA probe. However, the latter showed a stronger nuclear staining intensity.

The majority of the patients had serum antibodies to both EBV-VCA and EBV-nuclear antigens (EBNA). Interestingly enough, high serologic EBV titers were not related to the presence of EBV antigens in lymph node cells. Positive immunostaining of peripheral blood lymphocytes with anti-EBV antigens was detected in only 1 out of 15 HIV-positive patients with a frequency of about $1{:}10^4$ circulating lymphocytes. Peripheral blood lymphocytes were negative after in situ hybridization with EBV-DNA probe.

Discussion

Previous studies from this and other laboratories have demonstrated that activation and proliferation of B lymphocytes present in germinal centers represent one of the most characteristic and intriguing features of hyperplastic PGL [25]. It is also well documented that a marked polyclonal hypergammaglobulinemia is often associated to PGL and AIDS-related syndromes. However, it has not yet been established which stimulus might be responsible for B germinal cell activation and polyclonal immunoglobulin production. EBV has been found associated with reactive benign B cell proliferations [26] and some authors [27] have suggested that EBV, reactivated by HIV, could be responsible for the germinal center cell activation followed by polyclonal hypergammaglobulinemia. In the present paper we asked whether these features might be reflected by an increased number of EBV-positive B cells at lymphoid tissue level. Actually, during EBV-induced mononucleosis numerous circulating B lymphocytes are EBV-positive [28–30]. The present results demonstrate that in HIV-positive patients affected by PGL, although the majority of them had significant anti-EBV titers, EBV-VCA and EBV-EA-positive cells were observed in lymphoid tissues with a very low frequency; in addition, we have shown that in such patients EBV genome-positive cells were never detected either in in situ or in Southern blot hybridization. Furthermore, cells showing integrated EBV genome were never observed in peripheral blood lymphomononuclear cells. These negative results cannot be due to the methods used since a substantial proportion of cells of the control viral producer lines expressed both EBV antigens and EBV-integrated genome.

The absence of cells infected by EBV and expressing EBV antigens and genome in lymph nodes of HIV-positive patients may indicate that EBV is not directly involved in the pathogenesis of germinal center B cell polyclonal activation observed in PGL. Therefore, we must postulate the existence of other mechanisms. It may be suggested a possible role of HIV or HIV antigens in the induction of germinal center activation perhaps through the action of HIV immunocomplexes presented to B germinal center cells through the accessory cells [19]. Autoimmune mechanisms could also contribute to the development of PGL. Actually, a homology between two sequences present in the B1 domain of HLA class II molecules and in the gp41 region of HIV envelop protein has recently been indentified [31]. The same authors have also reported that approximately one third of AIDS patients had circulating antibodies recognizing self class II antigens. Autoimmune mechanisms explaining the B cell activation and proliferation can also be supported by a recent finding [32] demonstrating that patients with AIDS and AIDS-related conditions produce an IgG autoantibody reacting with HIV-infected $CD4^+$ T lymphocytes. However, the possibility that EBV might be, at least in part, involved in the pathogenesis of PGL cannot be completeley ruled out; actually EBV could be expressed, to such a low level, to infect only very few cells with few viral copies per cell. Thus, EBV antigens and genome are not detectable at tissue level using immunohistology and hybridization techniques.

References

1 Pope, J. H.: Establishment of cell lines from peripheral leukocytes in infectious mononucleosis. Nature *216:* 810–811 (1967).

2 Pope, J. H.; Horne, M. K.; Scott, W.: Transformation of fetal leukocytes in vitro by filtrates of a human leukemic cell line containing herpes-like virus. Int. J. Cancer *3:* 857–866 (1968).

3 Klein, E.; Masucci, M. G.: Cell mediated immunity against Epstein-Barr virus infected B lymphocytes. Springer Semin. Immunopathol. *5:* 63–73 (1982).

4 Wallace, L. E.; Rickinson, A. B.; Rowe, M.: Epstein-Barr virus-specific cytotoxic T-cell clones restricted through a single HLA antigen. Nature *297:* 413–415 (1982).

5 Blazar, B. A.; Patarroyo, M.; Klein, E.; Klein, G.: Increased sensibility of human lymphoid lines to natural killer cells after induction of the Epstein-Barr viral cycle by superinfection or sodium butyrate. J. Exp. Med. *151:* 614–627 (1980).

6 Pearson, G. R.; Orr, T. W.: Antibody-dependent lymphocyte cytotoxicity against cells experiencing Epstein-Barr virus antigens. JNCI *50:* 485–488 (1986).

7 Hanto, D. W.; Frizzera, G.; Gajl-Peczalska, K. J.; Sakamoto, K.; Purtilo, D. T.; Balfour, M. M., Jr.; Simmons, R. L.; Najarian, J. S.: Epstein-Barr virus-induced B cell lymphoma after renal transplantation: Acyclovir therapy and transition from polyclonal to monoclonal B-cell proliferation. N. Engl. J. Med. *306:* 913–918 (1982).

8 Shirley, L. R.; Buckley, R. H.; Borowitz, M. J.; Welt, S. I.; Rostyn, D. D.: Monoclonal

immunoglobulin secreting lymphoma in a patient with severe combined immunodeficiency disease. Clin. Exp. Immunol. *48:* 666–674 (1982).

9 Reece, E. R.; Gartner, J. G.; Seamayer, T. A.; Joucas, J. H.; Pagano, J. S.: Epstein-Barr virus in a malignant lymphoproliferative disorder of B-cells occurring after thymic epithelial transplantation for combined immunodeficiency. Cancer Res. *41:* 4243–4247 (1981).

10 Saemundsen, A. K.; Berkel, A. I.: Henle, W.; Henle, G.; Anvret, M.; Sanal, D.; Ersoy, F.; Caglar, M.; Klein, G.: Epstein-Barr virus carrying lymphoma in a patient with ataxia-teleangiectasia. Br. J. Med. *282:* 425–427 (1982).

11 Penn, I.: Malignant lymphomas in organ transplant recipients. Transplant Proc. *13:* 736–738 (1981).

12 Hanto, D. W.; Gajl-Peczelska, K. J.; Frizzera, G.; Arthur, D. C.; Balfour, M. M., Jr.; McClain, K.; Simmons, R. L.; Najarian, J. S.: Epstein-Barr virus (EBV) induced polyclonal and monoclonal B cell lymphoproliferative diseases occurring after renal transplantation. Ann. Surg. *198:* 356–369 (1983).

13 Pellicci, P. G.; Knowles, D. M., II; Arlin, Z. A.; Wieczorek, R.; Luciw, P.; Dina, D.; Basilico, C.; Dalla Favera, R.: Multiple monoclonal B cell expansions and c-myc oncogene rearrangements in acquired immune deficiency syndrome-related lymphoproliferative disorders. J. Exp. Med. *164:* 2049–2076 (1986).

14 Purtilo, T. D.: Opportunistic cancers in patients with immunodeficiency syndromes. Arch. Pathol. Lab. Med. *111:* 1123–1129 (1987).

15 Barre–Sinoussi, F.; Chermann, J. C.; Rey, F.; Nugeyre, M. T.; Chamaret, S.; Gruest, J.; Dauguet, C.; Axler-Blin, C.; Vezinet-Brun, F.; Rouzioux, C.; Rozenbaum, W.; Montagnier, L.: Isolation of a T lymphotropic retrovirus from a patient at risk for acquired immunodeficiency syndrome. Science *220:* 868–871 (1983).

16 Popovic, M.; Sarngadharan, M. G.; Read, E.; Gallo, R. C.: Detection isolation and continuous production of cytopathic retrovirus (HTLV-III) from patients with AIDS and pre-AIDS. Science *224:* 497–500 (1984).

17 Baroni, C. D.; Pezzella, F.; Mirolo, M.; Ruco, L. P.; Rossi, G. B.: Immunohistochemical demonstration of p24 HTLVIII major core protein in different cell types within lymph nodes from patients with lymphadenopathy syndrome (LAS). Histopathology *10:* 5–13 (1986).

18 Biberfeld, P.; Chayt, K. J.; Marselle, L. M.; Biberfeld, G.; Gallo, R. C.; Harper, M. E.; HTLV-III expression in infected lymph nodes and relevance to pathogenesis of lymphadenopathy. Am. J. Pathol. *125:* 436–442 (1986).

16 Tenner-Racz, K.; Racz, P.; Bofill, M.; Schulz-Meyer, A.; Dietrich, M.; Kern, P.; Weber, J.; Pinching, A. J.; Veronese-Di Marzo, F.; Popovic, M.; Klatzmann, D.; Gluckman, J. C.; Janossy, G.: HTLV-III/LAV viral antigens in lymph nodes of homosexual men with persistent generalized lymphadenopathy and AIDS. Am. J. Pathol. *126:* 9–15 (1986).

20 Pezzella, M.; Pezzella, F.; Galli, C.; Macchi, B.; Verani, P.; Sorice, F.; Baroni, C. D.: In situ hybridization of human immunodeficiency virus (HTLV-III) in cryostat sections of lymph nodes of lymphadenopathy syndrome patients. J. Med. Virol. *22:* 135–142 (1987).

21 Rechavi, G.; Ben-Bassati, I.; Benkowicz, M.; Martinowitz, U.; Brok-Simoni, F.; Neumann, Y.; Vansover, A.; Gotlieb-Stematsky, T.; Ramot, B.: Molecular analysis of Burkitt's leukemia in two hemophilic brothers with AIDS. Blood *70:* 1713–1717 (1987).

22 Kishishita, M.; Luka, J.; Vrohan, B.; Poduslo, J.; Pearson, G. R.: Production of mono-

clonal antibody to a late intracellular EBV-induced antigen. Virology *133:* 363–375 (1984).

23 Luka, J.; Miller, G.; Jornvall, H.; Pearson, G.R.: Characterization of the restricted component of EBV early antigen as a cytoplasmic filamentous protein. J. Virol. *58:* 745–756 (1986).

24 Di Marzo Veronese, F.; Sarngadharan, M.G.; Raman, R.; Markham, P.D.; Popovic, M.; Booner, A.J.; Gallo, R.C.: Monoclonal antibody specific for p24, the major core protein for HTLVIII. Proc. Natl. Acad. Sci. USA *82:* 5199–5202 (1985).

25 Baroni, C.D.; Pezzella, F.; Stoppacciaro, A.; Mirolo, M.; Pescarmona, E.; Vitolo, D.; Cassano, A.M.; Barsotti, P.; Nicoletti, L.; Ruco, L.P.; Uccini, S.: Systemic lymphadenopathy (LAS) in intravenous drug abusers. Histology, immunohistochemistry and electron microscopy: pathogenetic correlations. Histopathology *9:* 1275–1293 (1985).

26 Henle, G.; Henle, W.; Diehl, V.: Relation of Burkitt's tumor associated herpes-type virus to infectious mononucleosis. Proc. Natl. Acad. Sci. USA *59:* 94–101 (1968).

27 Ragona, G.; Sirianni, M.C.; Soddu, S.; Vercelli, B.; Sebastiani, G.; Piccoli, M.; Aiuti, F.: Evidence for dysregulation in the control of Epstein-Barr virus latency in patients with AIDS-related complex. Clin. Exp. Immunol. *66:* 17 (1986).

28 Rocchi, G.; De Felici, A.; Ragona, G.; Heints, A.: Quantitative evaluation of Epstein-Barr virus-infected mononuclear peripheral blood leukocytes in infectious mononucleosis. N. Engl. J. Med. *296:* 132–134 (1977).

29 Robinson, J.; Smith, D.; Niederman, J.: Mitotic EBNA-positive lymphocytes in peripheral blood during infectious mononucleosis. Nature *287:* 334–335 (1980).

30 Robinson, J.F.; Smith, D.; Niederman, J.: Plasmacytic differentiation of circulating Epstein-Barr virus-infected B lymphocytes during acute infectious mononucleosis. J. Exp. Med. *153:* 235–244 (1981).

31 Golding, H.; Robey, F.A.; Gates F.T., III; Linder, W.; Beining, P.R.; Hoffman, T.; Golding, B.: Identification of homologous regions in human immunodeficiency virus 1 gp41 and human MHC class II B1 domain. 1. Monoclonal antibodies against the gp41-derived peptide and patients' sera react with native HLA class II antigens, suggesting a role for autoimmunity in the pathogenesis of acquired immune deficiency syndrome. J. Exp. Med. *167:* 914–923 (1988).

32 Stricker, R.B.; McHugh, T.M.; Moody, D.J.; Morrow, W.J.W.; Stites, D.P.; Shuman, M.A.; Levy, J.A.: An AIDS-related cytotoxic autoantibody reacts with a specific antigen on stimulated $CD4^+$ T cells. Nature *327:* 710–713 (1987).

Carlo D. Baroni, MD, Second Chair of Pathological Anatomy,
Viale Regina Elenà 324, I–00161 Rome (Italy)

Racz P, Haase AT, Gluckman JC (eds): Modern Pathology of AIDS and Other Retroviral Infections. Basel, Karger, 1990, pp 110–129

AIDS-Related Central Nervous System Lymphoma

Demonstration of Epstein-Barr Virus DNA by in situ Hybridisation[1]

Stephen J. Hamilton-Dutoit[a], *Jens Karkov*[b], *Maria B. Franzmann*[c], *Gorm Pallesen*[a]

[a]Laboratory of Immunohistology, University Institute of Pathology, Kommunehospitalet, Aarhus; [b]Department of Pathology, Rigshopitalet, Copenhagen; [c]Department of Pathology, Hvidovre Hospital, Denmark

Malignant lymphoma occurs with increased frequency in individuals infected with human immunodeficiency virus (HIV) [1, 2]. Estimates of the incidence of AIDS-related lymphomas vary, but the cumulative rate amongst Danish AIDS patients is currently 7.5% [3]. Whilst involvement of the central nervous system (CNS) is rare in lymphomas arising in immunocompetent individuals (accounting for less than 2% of such cases [4]) it is a common feature of AIDS-related lymphoma [5]. Published reports suggest that, depending on the type of study involved, approximately one quarter of AIDS-related lymphomas present with primary CNS disease, and that the brain may be involved along with other organs in up to 45% of cases [1, 6–9]. The association of primary CNS lymphoma with HIV infection was recognised early in the AIDS epidemic [10, 11] and was considered diagnostic of AIDS. Although the Centers for Disease Control (CDC) case definition of AIDS was later revised to include the occurrence of all high grade lymphomas as a diagnostic criterion [12], the frequency with which AIDS-related lymphomas present with primary disease in the CNS remains a characteristic feature of these tumours [5].

The pathogenesis of AIDS-related lymphoma is not understood, but a possible role for Epstein-Barr virus (EBV) has been proposed based on (a) evidence of defective immune regulation of EBV infection in AIDS [13–17],

[1]This study was supported by the Danish Medical Research Council (AIDS 12-7938), the Lions Club of Denmark, and EEC contract MR4-0019/F.

and (b) clinical and pathological similarities observed between AIDS-related lymphoma and lymphomas arising in other immunodeficiency states, notably those known to be associated with EBV infection. The latter include both conditions which are of congenital origin (for example the X-linked lymphoproliferative syndrome [18]) and those with an iatrogenic aetiology (such as posttransplantation immunodeficiency states [19]).

In support of this proposal, chromosomal translocations similar to those seen in Burkitt's lymphoma have been described in AIDS-related lymphoma [20–25] and in some cases EBV nuclear-associated antigen [21–23, 25–27] has been identified in tumour cells. A number of studies have also provided evidence for the presence of EBV genomes within AIDS-related lymphoma tissues based upon the study of extracted DNA using nucleic acid hybridisation in Southern blot analysis [21–25, 27–29]. Such filter-hybridisation methods suffer, however, from the disadvantage that tissue morphology and architectural features are destroyed, and it is not therefore possible to be certain of the nature of the cells (i.e. tumour or reactive) containing viral sequences in a positive specimen. This is not a problem with in situ hybridisation. Furthermore, if viral DNA is confined to only a small proportion of cells in a tissue, then in situ hybridisation may be more sensitive at detecting these than Southern blotting. Finally, although techniques for performing filter hybridisation using DNA extracted from paraffin blocks are being developed, most studies using Southern blotting continue to require fresh frozen tissue specimens thus restricting the number of cases which may be investigated. Methods are, however, now available for the detection of EBV-DNA using in situ hybridisation on routinely processed paraffin-embedded tissues, making much more extensive (and retrospective) studies possible.

We examined 6 cases of AIDS-related lymphoma with CNS involvement arising in Danish patients for the presence of EBV-DNA using a sensitive in situ hybridisation technique. In 4 cases we were able to identify EBV genomes directly in tumour cells.

Materials and Methods

Tissues

Tissues from 6 Danish patients (5 homosexuals and 1 haemophiliac) with AIDS-related malignant lymphoma were selected from the files of the University Institute of Pathology, Aarhus Kommunehospital, and from the University Departments of Pathology of Rigshospitalet and Hvidovre Hospital, Copenhagen. All specimens were obtained at autopsy. Four cases presented with primary CNS disease, whilst in 2 cases there was involvement of other

Table 1. Clinical features of AIDS-related CNS lymphomas

Patient No.	Sex	Age	Risk group	Site of lymphoma	Date (month/year)			
					HIV diagnosis	onset CNS symptoms	lymphoma diagnosis	death
1	M	24	haemophilia A	brain and lung	04/85	12/84	at autopsy	05/87
2	M	42	homosexual	disseminated	09/85	11/87	at autopsy	12/87
3	M	36	homosexual	primary CNS	06/88	05/88	at autopsy	06/88
4	M	46	homosexual	primary CNS	07/85	05/85	at autopsy	08/85
5	M	33	homosexual	primary CNS	01/85	01/87	at autopsy	02/87
6	M	32	homosexual	primary CNS	12/81	10/82	at autopsy	01/83

organs at post mortem. One of the latter (case 1) presented clinically with symptoms localised to the CNS before developing evidence of lung involvement. Clinical details of the patients are shown in table 1. Cases were classified histologically according to the updated Kiel classification [30]. All tissues were fixed in buffered formalin and embedded in paraffin using standard histological procedures. Sections of oral hairy leukoplakia from a patient with AIDS, and of a tonsil from a patient with acute infectious mononucleosis were used as positive control tissues for EBV in situ hybridisation.

Plasmids and Probes

Plasmid pBa-W (*Bam*HI-W-internal repetitive fragment of EBV strain M-ABA, subcloned in pBR322) [31] was provided by Dr. G. W. Bornkamm, Freiburg, FRG. pCM5018 (human CMV strain AD169 *Eco*RI-J-fragment subcloned into pACYC184) [32] was a gift from Dr. B. Fleckenstein, Erlangen, FRG. Both plasmids were kindly prepared by Dr. H. Herbst, West Berlin. Total plasmid DNA was labelled with ^{35}S-dCTP (NEN, specific activity 1,200 Ci/mmol) by nick translation to a specific activity of at least 1×10^9 dpm/μg [33]. ^{35}S-labelled pBR322 DNA hybridisation probe was purchased from NEN.

In situ Hybridisation

In situ hybridisation studies were performed on formalin-fixed tissue embedded in paraffin, using a modified procedure based upon previous described methods [34, 35]. Tissue sections were placed on glass slides pretreated with 3-aminopropyltriethoxysilane (Merck, Munich, FRG) to prevent section loss [36], baked overnight at 65 °C, and deparaffinised with xylene. The prehybridisation procedure consisted of sequential incubation of slides at room temperature in 0.2 *N* HCl (10 min) and 0.01% Triton X-100 (90 s) with intervening washes in 2 × SSC (0.3 *M* sodium chloride and 0.03 *M* sodium citrate, pH 7.6).

Sections were subjected to pronase digestion (Boehringer Mannheim, FRG) at a concentration of 1 mg/ml for 10 min at 37 °C, washed again in 2 × SSC, and then acetylated with 0.25% acetic anhydride in 0.1 *M* triethanolamine (pH 8.0) for 10 min, followed by dehydration through graded alcohols and air drying. The hybridisation mixture consisted of 40 ng/ml

Table 2. Monoclonal antibodies

Antibody	Source	Specificity (major)	Ref.
4KB5	Dr. D.Y. Mason	B cell	48
LN1	Clonab	B cell	49
L26	Dakopatts	B cell	50
UCHL1	Dr. P.C.L. Beverley	T cell	51
MT1	Bio-Nuclear Services	T cell	52

labelled probe, 0.25 mg/ml sheared salmon sperm DNA, 50% deionised formamide, 2 × SSC, 10% dextran sulphate, and 100 m*M* dithiothreitol. Some 25 μl of hybridisation mixture were applied to an average-sized section. Slides were covered with siliconised coverslips and incubated at 90 °C on a heating block (Enzo, New York) for 6 min. After overnight hybridisation at 37 °C, coverslips were removed in 50% formamide/0.1 × SSC at 37 °C and then washed in 50% formamide/0.1 × SSC at 37 °C for 4 h. Slides were then washed for 30 min each in 2 × SSC followed by 0.1 × SSC, both at room temperature. The sections were dehydrated in graded alcohols containing 0.3 *M* ammonium acetate and air dried, before being dipped in Ilford G5 emulsion (with 0.3 *M* ammonium acetate), exposed at 4 °C for 2–8 days, developed in Kodak D19, fixed with Kodak rapid fixer, and counterstained with haematoxylin.

Cells were regarded as positive for viral DNA if they showed deposition of grains clearly in excess of the background level seen over random cells. Quantitative grain counts per cell were not performed.

Controls

In all experiments the ^{35}S-radiolabelled CMV probe and the ^{35}S-radiolabelled pBR322 DNA probe (without the EBV insert) were used on adjacent sections as negative probe controls. Sections of oral hairy leukoplakia and acute infectious mononucleosis tonsils were included in each batch of experiments as positive tissue controls. Non-hybridised sections and plain glass objective slides were included in each autoradiography batch as controls of background levels of signal.

Immunohistology

Paraffin sections were stained with the monoclonal antibodies shown in table 2 using a standard 3-stage immunoperoxidase technique. Sections of lymph node or tonsil showing non-specific hyperplasia were used as positive controls.

Results

The results of histologic typing, immunohistology and in situ hybridisation are given in tables 3 and 4. All cases were of high grade morphology according to the Kiel classification. Five of the cases were immunoblastic

Table 3. AIDS-related CNS lymphoma: results of immunophenotyping with monoclonal antibodies[1]

Patient No.	Monoclonal antibodies					Immunophenotype
	4KB5	LN1	L26	UCHL1	MT1	
1	–	2+	3+	1+	–	B cell
2	–	1+	2+	1+	–	B cell
3	–	1+	2+	–	–	B cell
4	–	–	3+	2+	–	B cell
5	–	2+	2+	3+	–	B cell
6	–	2+	3+	–	–	B cell

[1] Slight/moderate/strong positive staining = 1+/2+/3+.

Table 4. AIDS-related CNS lymphoma: histological and immunohistological type and results of examination for EBV genomes using in-situ hybridisation

Patient No.	Histological diagnosis	Immunophenotype	EBV genomes present
1	immunoblastic (monomorphic)	B cell	yes
2	lymphoblastic (non-Burkitt)	B cell	no
3	immunoblastic (PC differentiation)	B cell	no
4	immunoblastic (PC differentiation)	B cell	yes
5	immunoblastic (PC differentiation)	B cell	yes
6	immunoblastic (PC differentiation)	B cell	yes

PC = plasmocytic differentiation. However, clonality of the plasma cells could not be established.

lymphomas. One of these was relatively monomorphic and was similar to lymphomas arising in immunocompetent individuals. The other 4 cases, however, exhibited a number of histological features which made categorisation using conventional morphological criteria rather difficult. Chief amongst these 'aberrant' characteristics was a significant degree of polymorphism of

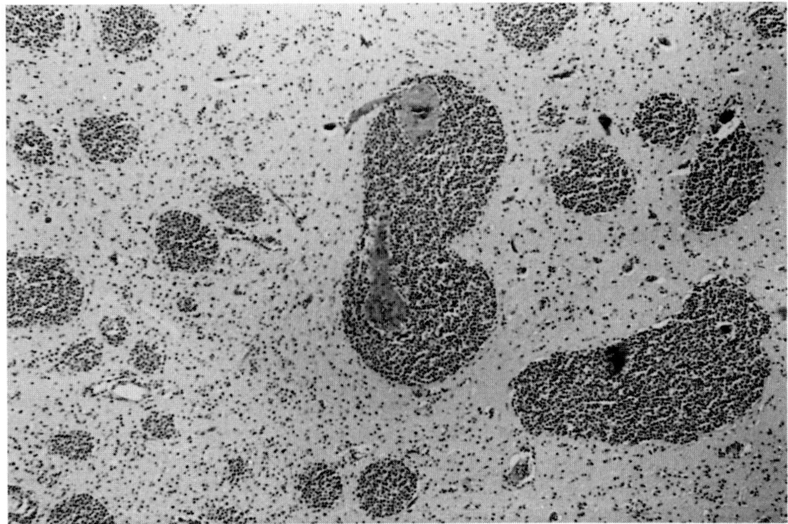

Fig. 1. Section of brain from a 24-year-old haemophiliac with AIDS-related non-Hodgkin's lymphoma (case 1). Tumour cells surround blood vessels and infiltrate diffusely through the adjacent tissue. HE. × 50.

the lymphocytes. This feature has been noted before by Frizzera et al. [37] in lymphomas arising in posttransplantation patients and the tumours designated 'polymorphic diffuse B cell lymphoma'. These 4 cases all included quite numerous plasma cells, although it proved difficult to be sure whether these represented plasmacytic differentiation of the tumour, or were part of the reactive cell population. Finally, one tumour was a lymphoblastic lymphoma of non-Burkitt type.

All six lymphomas had a B cell phenotype in paraffin section. An unusual feature (which we have also seen in other high grade B cell lymphomas, particularly those arising in immunodeficient patients) was the labelling of tumour cells with monoclonal antibody UCHL1 in 4 of the 6 cases.

In situ hybridisation with the ^{35}S-radiolabelled *Bam*HI-W EBV probe revealed a clear positive autoradiographic signal over the nuclei of tumour cells in 4 cases (cases 1, 4, 5, 6; fig. 1–8). These positive cells varied in size, but were most often large and immunoblast-like. In some cases positive cells were relatively few in number whilst in others each high power field contained many such cells (fig. 2b). In contrast, the surrounding tissue and non-tumour cell population showed no hybridisation signal over background levels. In the

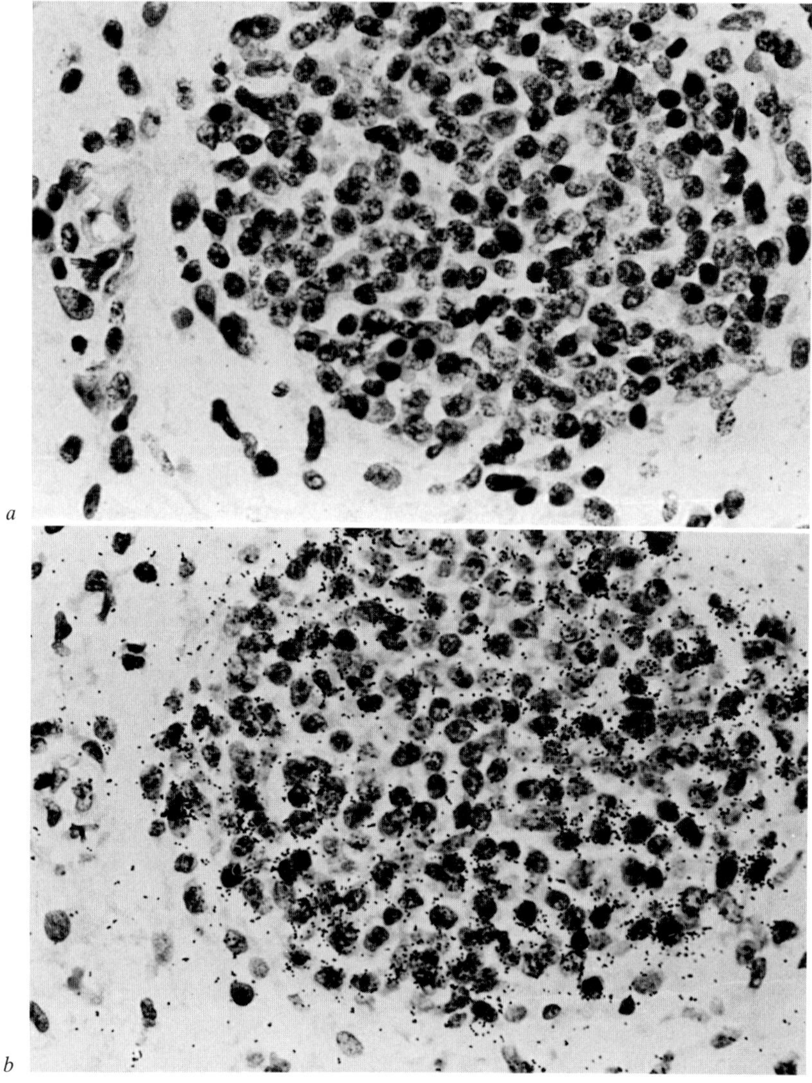

Fig. 2. Sections of brain from the same patient as in figure 1 (case 1) showing a nodule of lymphoma cells associated with a blood vessel. *a* In situ hybridisation with the control ^{35}S-labelled CMV probe shows only minimal background autoradiographic signal. *b* An adjacent section hybridised with the EBV-specific ^{35}S-labelled probe shows a strong signal over almost all the tumour cells. Haematoxylin counterstain. *a, b* × 490.

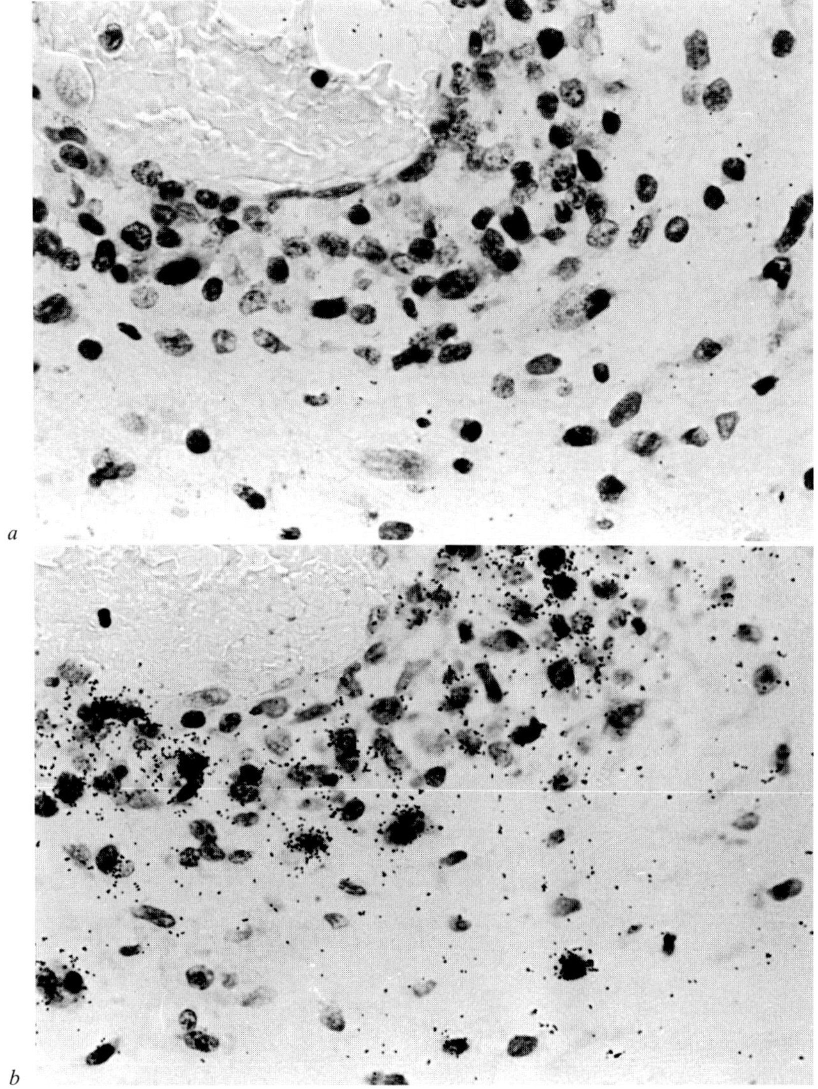

Fig. 3. A second field from the same sections shown in figure 2 (case 1). Scattered tumour cells are seen around a blood vessel. *a* In situ hybridisation with the control CMV probe shows only background signal. *b* In an adjacent section in situ hybridisation with the EBV probe reveals a strong positive signal over lymphoma cells. There is no labelling above background of normal parenchymal cells. Haematoxylin. *a, b* × 490.

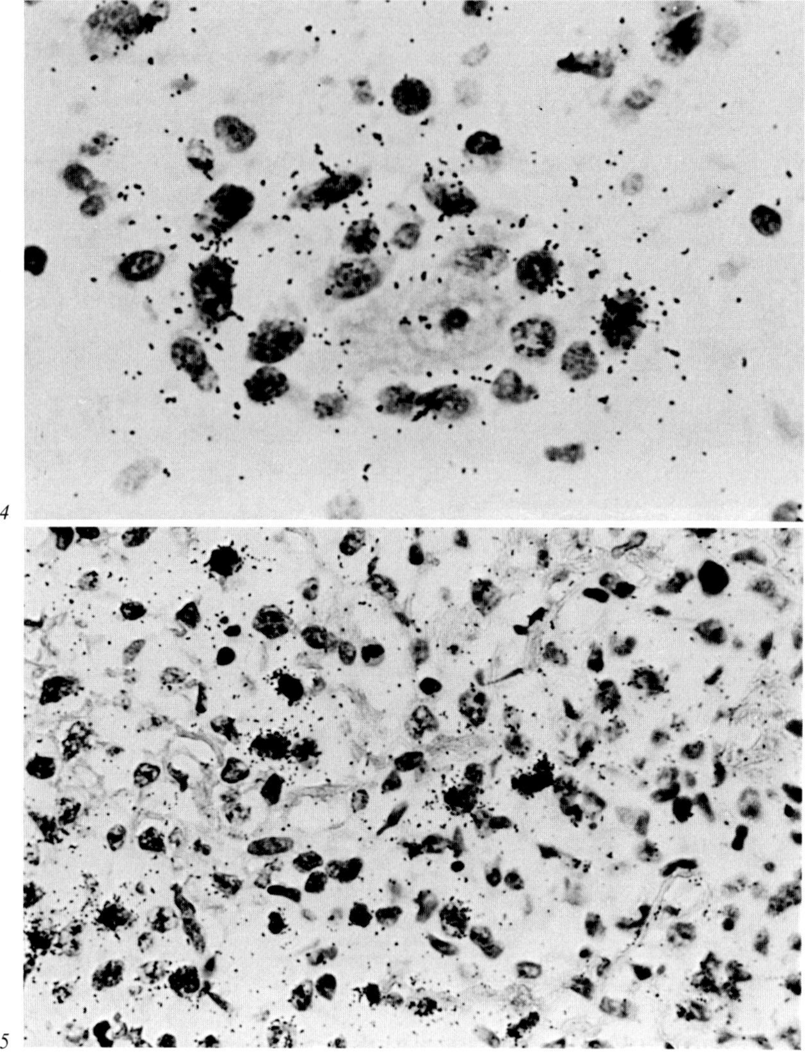

Fig. 4. Cerebral cortex from case 1. In situ hybridisation with the EBV-specific probe gives positive labelling of lymphoma blast cells. Near the centre of the field a neuron can be seen which shows only a background level of signal. Haematoxylin counterstain. × 740.

Fig. 5. A section of lung infiltrated by lymphoma from the same patient as in the previous figures (case 1). In situ hybridisation with the EBV probe shows strong labelling of the lymphoma cells similar to that seen in the brain (fig. 2–4). Lung parenchymal cells are negative. Haematoxylin. × 490.

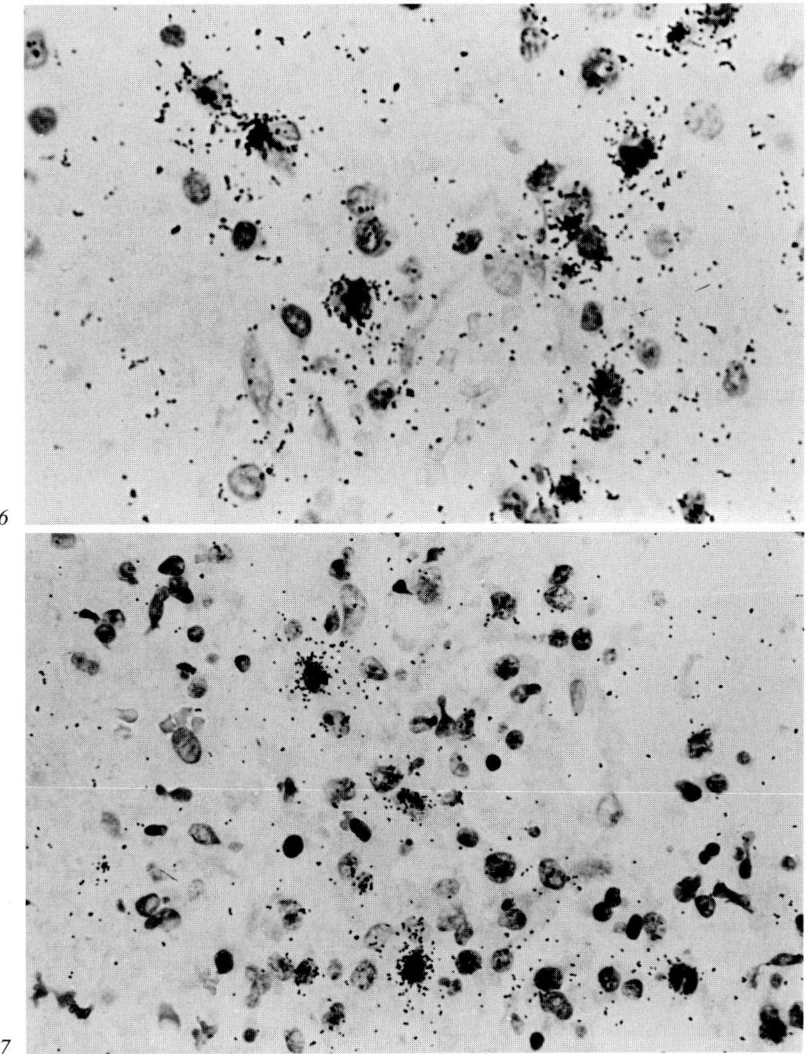

6

7

Fig. 6. Cerebral cortex from a 46-year-old homosexual with primary AIDS-related CNS lymphoma (case 4). In situ hybridisation for EBV. After 7 days' autoradiographic exposure there is a slight to moderate background signal, but very dense labelling is present over lymphoma cells. Haematoxylin counterstain. × 740.

Fig. 7. Section of brain from a 33-year-old homosexual with primary AIDS-related CNS lymphoma (case 5). In situ hybridisation for EBV reveals scattered strongly positive lymphoma cells after 6 days' autoradiographic exposure. Haematoxylin counterstain. × 490.

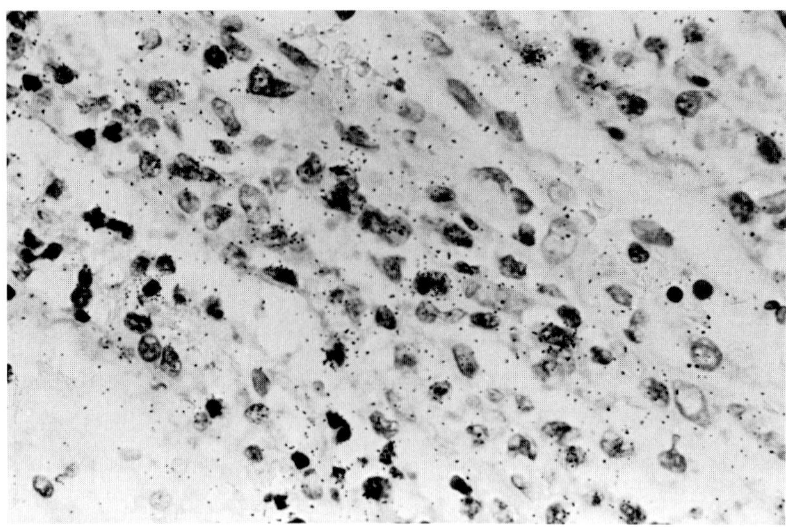

Fig. 8. Primary AIDS-related CNS lymphoma in a 32-year-old homosexual (case 6). In situ hybridisation for EBV reveals positively labelled tumour cells in and around a blood vessel. Haematoxylin counterstain. × 490.

remaining 2 patients (cases 2 and 3) no specific hybridisation with the EBV-DNA probe was seen.

Of the 4 cases presenting with primary CNS lymphoma, 3 were found to contain EBV-DNA. The other patient positive for EBV-DNA in the brain (case 1) also initially presented with symptoms localised to the CNS, but later developed lung infiltrates and was found to have lymphoma in both lungs and brain at autopsy and so was not classified as a primary CNS lymphoma. Sections from lung and brain in this case were examined with the EBV probe, and identical patterns of strong specific hybridisation over many tumour cells were found in both tissues (fig. 2–5).

No signal over background was detected in any of the EBV-positive cases when the ^{35}S-radiolabelled probes for CMV and pBR322 were substituted in place of the EBV probe. The signal found in EBV-positive AIDS-related lymphoma tumour cells was comparable with, but generally less strong than that seen in EBV-positive lymphoid blasts in the tonsils from cases of acute infectious mononucleosis used as one of the positive tissue controls (fig. 9). In

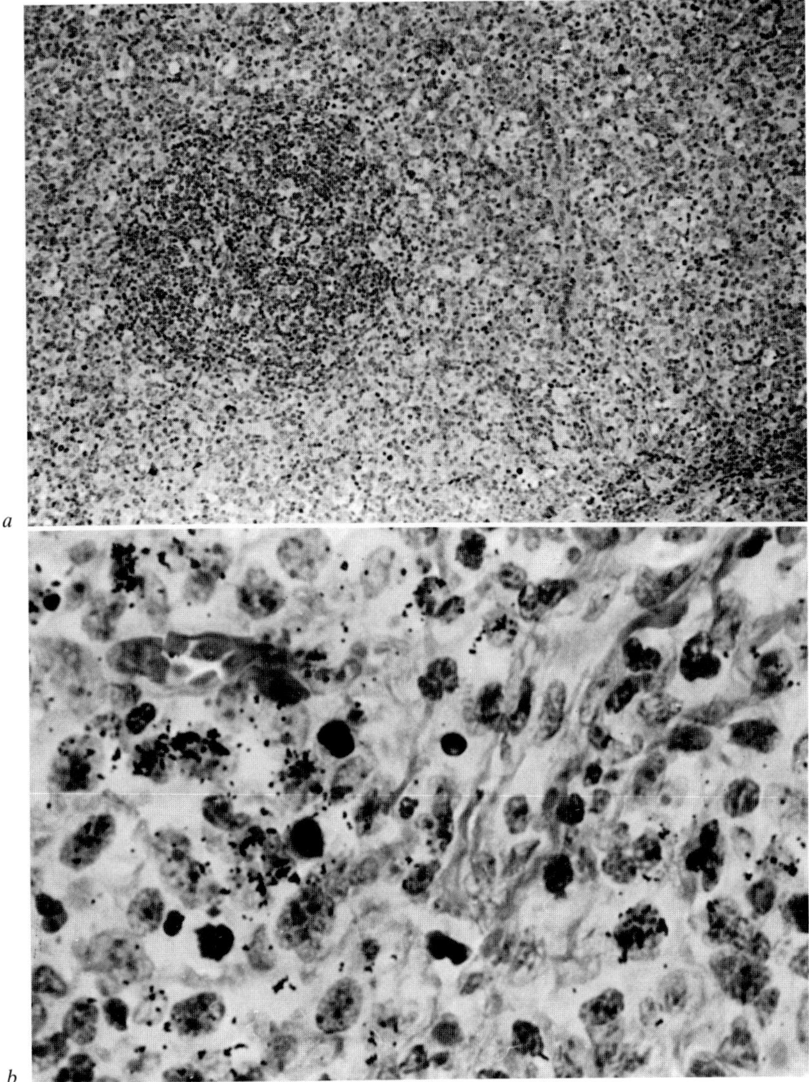

Fig. 9. Tonsil from a patient with acute infectious mononucleosis used as a positive control tissue for in situ hybridisation employing a ^{35}S-labelled probe specific for EBV. *a* A follicle surrounded by an expanded paracortical area containing many large transformed lymphoid blast cells. *b* At higher power many of these paracortical blast cells show strong positive autoradiographic labelling indicating the presence of EBV genomes. Haematoxylin counterstain. *a* × 125. *b* × 740.

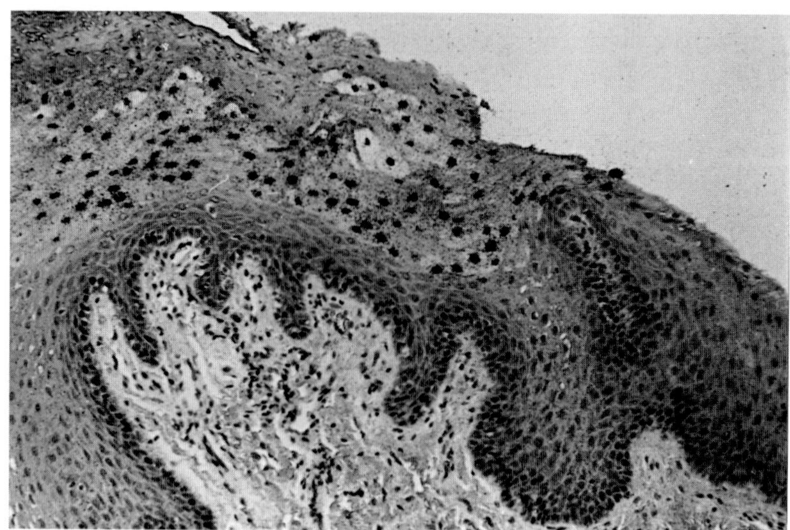

Fig. 10. Section of tongue from an AIDS patient with oral hairy leukoplakia used as a positive control tissue for in situ hybridisation employing a ^{35}S-labelled probe specific for EBV. A dense autoradiographic signal is seen covering epithelial cell nuclei after only 2 days' exposure indicating a very high number of EBV genomes. Haematoxylin counterstain. × 125.

contrast, a very much greater signal was detected over epithelial cell nuclei of control sections of oral hairy leukoplakia after 2 days' autoradiography (fig. 10).

Discussion

We have used in situ hybridisation to demonstrate directly the presence of EBV nucleic acids in tumour cells of a substantial proportion (4 out of 6 cases) of AIDS-related lymphoma with involvement of the CNS. We have not calculated the absolute copy number that it is possible to detect with our technique, but some estimate of this can be gained by comparing the results seen in the lymphomas with those in our positive control tissues. In contrast to the strong nuclear signal detected in the positive control sections of oral hairy leukoplakia (an epithelial lesion in which EBV replicative infection is present associated with some 200 viral copies per cell [38]) the signal over AIDS-related lymphoma cells was relatively weak suggesting the presence of a low viral copy number and a latent infection. In keeping with this was the comparable signal seen in control sections of tonsil from acute infectious mononucleosis. We have recently shown [35] that EBV-DNA can be consistently detected in the transformed lymphoid blasts found in this disorder and there is again evidence suggesting that this is indicative of a latent rather than replicative infection.

All cases in our study were B cell lymphomas. Five were classified as immunoblastic lymphomas, including all the tumours positive for EBV genomes, indicating that association with EBV is not confined to tumours with the classic morphological features of Burkitt's lymphoma. Four of these immunoblastic lymphomas showed an unusual degree of polymorphism similar to that described by Frizzera et al. [37] in posttransplantation lymphomas and designated 'polymorphic diffuse B cell lymphoma'. It has been suggested that this feature may indicate those tumours which are polyclonal [39]. Unfortunately, fresh tumour tissue was not available to us from our cases and so we were unable to perform studies to establish clonality of the lesions.

Evidence that EBV may play a role in the pathogenesis of AIDS-related lymphoma has come from various sources. Lymphomas arise with increased frequency in immunocompromised patients, including those suffering from inherited immunodeficiency states [40], iatrogenic immunodeficiencies (such as immunosuppressed organ transplant patients) [39, 41] and African children at risk of developing Burkitt's lymphoma [42]. EBV has been implicated in the pathogenesis of these tumours [43]. These immunodeficiency associated lymphomas show many clinical and pathological similarities to AIDS-related lymphomas (including their predilection for CNS involvement), suggesting that their causes may also be similar. In support of this, EBV-nuclear associated antigen has been demonstrated in AIDS-related lymphoma cells [21–23, 25–27], and characteristic chromosomal translocations [20–25] and c-*myc* oncogene rearrangements and activation [25] have been described which resemble those classically seen in Burkitt's lymphoma. There is also evidence that immune control of EBV infection is dysregulated in HIV carriers [29]. In particular, these individuals have increased levels of EBV-DNA in their saliva [16], increased anti-EBV antibody titres [13–15], increased levels of circulating B cells [13] and impaired T cell mediated control of EBV-infected lymphocytes [13]. In addition, in oral hairy leukoplakia the epithelial cells of the tongue support a replicative EBV-infective cycle. This condition has only ever been described in association with HIV infection.

EBV genomes have previously been found within AIDS-related lymphoma tumour tissues using Southern blot analysis of extracted nucleic acid rather than in situ hybridisation [21–25, 27–29]. Reported cases have been limited in number and there have been few large series making it difficult to estimate the percentage of AIDS-related lymphoma positive for EBV-DNA by Southern blotting. In a review of published cases, Ernberg and Altiok [29] identified 29 tumours which had been studied for EBV, 58% of which were positive for either EBV antigens and/or genomes. In addition, they described their own series of 25 cases of AIDS-related lymphoma studied for EBV-DNA by Southern blotting, of which 7 (28%) were positive. This figure is comparable with the results found in a study by Subar et al. [25] who found EBV genomes in 6 (38%) of 16 cases of AIDS-related lymphoma by Southern blot analysis.

There would, therefore, appear to be a discrepancy between the percentage of AIDS-related lymphomas found to be positive for EBV by Southern blotting as compared with in situ hybridisation. There are a number of possible explanations for this. Techniques which rely upon examination of extracted nucleic acids are liable to sampling errors as it is not possible to be certain which cells are contained in the specimen for analysis. Thus necrotic or non-tumour tissue may be present leading to an underestimate of the number of positive cases. In situ hybridisation methods have the great advantage that tissue structure is preserved and the nature of the cells being examined can be identified. In addition, in situ hybridisation is potentially more sensitive than Southern blotting in detecting virally infected cells if the latter are only found in small numbers. In this case the viral copies present may be 'diluted' by the many negative cells during nucleic acid extraction until they are too few to be detected. One possible source of error in our present study is that it is based on a rather small number of cases. We have, however, recently published our preliminary results from a larger series of AIDS-related lymphomas (including tumours without CNS involvement) examined using in situ hybridisation [3]. In this study we found EBV genomes in 8 (50%) out of 16 cases, a figure which is also greater than that predicted from reports based upon the use of Southern blotting and which again suggests that in situ hybridisation may be more sensitive at detecting the virus. In addition, this latter series suggests that EBV may be found rather more frequently in tumours with CNS involvement (particularly those with primary CNS disease) as compared with those with exclusively systemic tumour (although the numbers of cases examined are still too few to draw firm conclusions). If such a discrepancy does exist it may reflect either an absolute difference in frequency of association with EBV, or a difference in the viral copy number per cell.

The explanation for the extraordinary propensity of immunodeficiency-associated lymphomas to give rise to CNS involvement remains unknown. It has, however, been speculated that this reflects the status of the CNS as a 'sanctuary site' in which normal mechanisms of immune surveillance may not be operative [41, 44]. In keeping with this is the observation that EBV-induced lymphoblastoid cell lines which fail to grow when transplanted subcutaneously into nude mice, will grow uncontrollably if placed in the mouse brain [45, 46]. In addition, EBV genomes have been detected by Southern blotting in a primary CNS lymphoma rising in a patient without evidence of immunosuppression or HIV infection [47], suggesting that the virus may be able to escape normal immunological control and induce lymphoma formation at this site in otherwise healthy individuals. There is evidence that patients with AIDS-related primary CNS lymphoma are particularly severely immunocompromised as compared with other AIDS patients [7, 8], a factor which may make them especially susceptible to the development of a virally induced tumour.

The demonstration of EBV in tumour tissues of AIDS-related lymphoma does not provide absolute proof that the virus plays a primary role in their pathogenesis, even when it can be established by in situ hybridisation that the virus is present within actual tumour cells. More substantial evidence for such a role is, however, provided by our findings in case 1, in which the virus was detected in tumour cells in both brain and lungs, but not in non-tumour tissues. This would suggest that the EBV is specifically associated with the tumour cells and is not merely a passenger.

More extensive studies are required to further examine the incidence of EBV in AIDS-related CNS lymphoma. In the present investigation we have been able to detect EBV-DNA using in situ hybridisation in tumour specimens obtained at autopsy which had been routinely processed and stored for up to 6 years prior to examination. These methods make possible extensive retrospective studies based on archive material. The development of in situ hybridisation techniques, such as the one described here, has provided pathologists with an ideal tool for investigating the role of EBV and other viruses in the pathogenesis of lymphoproliferations in both immunocompetent and immunodeficient individuals.

References

1 Ziegler, J.; Beckstead, J.; Volberding, P.; Abrams, D.; Levine, A.; Lukes, R.; Gill, P.; Burkes, R.; Meyer, P.; Metroka, C.; Mouradian, J.; Moore, A.; Riggs, S.; Butler, J.; Cabanillas, F.; Hersh, E.; Newell, G.; Laubenstein, L.; Knowles, D.; Odajnyk, C.; Raphael, B.; Koziner, B.; Urmacher, C.; Clarkson, B.: Non-Hodgkin's lymphoma in 90

homosexual men. Relation to generalized lymphadenopathy and the acquired immunodeficiency syndrome. N. Engl. J. Med. *311:* 565–570 (1984).
2 Levine, A.; Meyer, P.; Begandy, M.; Parker, J.; Taylor, C.; Irwin, L.; Lukes, R.: Development of B-cell lymphoma in homosexual men. Clinical and immunological findings. Ann. Intern. Med. *100:* 7–13 (1984).
3 Hamilton-Dutoit, S.; Pallesen, G.; Karkov, J.; Skinhøj, P.; Franzmann, M.; Pedersen, C.: Identification of EBV-DNA in tumour cells of AIDS-related lymphomas by in situ hybridisation. Lancet *i:* 554–555 (1989).
4 Henry, J.; Heffner, R.; Dillard, S.; Earle, K.; Davis, R.: Primary malignant lymphomas of the central nervous system. Cancer *34:* 1293–1302 (1974).
5 Gill, P.; Levine, A.; Meyer, P.; Boswell, W.; Burkes, R.; Parker, J.; Hofman, F.; Dworsky, R.; Lukes, R.: Primary central nervous system lymphoma in homosexual men. Clinical, immunologic, and pathologic features. Am. J. Med. *78:* 742–748 (1985).
6 Di Carlo, E.; Amberson, J.; Metroka, C.; Ballard, P.; Moore, A.; Mouradian, J.: Malignant lymphomas and the acquired immunodeficiency syndrome. Evaluation of 30 cases using a working formulation. Arch. Pathol. Lab. Med. *110:* 1012–1016 (1986).
7 Kalter, S.; Riggs, S.; Cabanillas, F.; Butler, J.; Hagemeister, F.; Mansell, P.; Newell, G.; Velasquez, W.; Salvador, P.; Barlogie, B.; Rios, A.; Hersh, E.: Aggressive non-Hodgkin's lymphomas in immunocompromised homosexual males. Blood *66:* 655–659 (1985).
8 Lowenthal, D.; Straus, D.; Wise Campbell, S.; Gold, J.; Clarkson, B.; Koziner, B.: AIDS-related lymphoid neoplasia. The Memorial Hospital experience. Cancer *61:* 2325–2337 (1988).
9 Levine, A.; Gill, P.; Meyer, P.; Burkes, R.; Ross, R.; Dworsky, R.; Krailo, M.; Parker, J.; Lukes, R.; Rasheed, S.: Retrovirus and malignant lymphoma in homosexual men. J. Am. Med. Ass. *254:* 1921–1925 (1985).
10 Snider, W.; Simpson, D.; Aronyk, K.; Nielsen, S.: Primary lymphoma of the nervous system associated with acquired immune-deficiency syndrome. N. Engl. J. Med. *308:* 45 (1983).
11 Case records of the Massachusetts General Hospital (Case 32-1983). N. Engl. J. Med. *309:* 359–369 (1983).
12 Centers for Disease Control: Revision of the case definition of acquired immunodeficiency syndrome for national reporting – United States. Morbid. Mortal. Weekly Rep. *34:* 373–375 (1985).
13 Birx, D.; Redfield, R.; Tosato, G.: Defective regulation of Epstein-Barr virus infection in patients with acquired immunodeficiency syndrome (AIDS) or AIDS-related disorders. N. Engl. J. Med. *314:* 874–879 (1986).
14 Ragona, G.; Sirianni, M.; Soddu, S.; Vercelli, B.; Piccoli, M.; Aiuti, F.: Evidence for disregulation in the control of Epstein-Barr virus latency in patients with AIDS-related complex. Clin. Exp. Immunol. *66:* 17–24 (1986).
15 Rinaldo, C.; Kingsley, L.; Lyter, D.; Rabin, B.; Atchison, R.; Bodner, A.; Weiss, S.; Saxinger, W.: Association of HTLV III with Epstein-Barr virus infection and abnormalities of T lymphocytes in homosexual men. J. Infect. Dis. *154:* 556–561 (1986).
16 Alsip, G.; Ench, Y.; Sumaya, C.; Boswell, R.: Increased Epstein-Barr virus DNA in oropharyngeal secretions from patients with AIDS, AIDS-related complex, or asymptomatic human immunodeficiency virus infections. J. Infect. Dis. *157:* 1072–1076 (1988).
17 Crawford, D.; Weller, I.; Iliescu, V.: Polyclonal activation of B cells in homosexual men. N. Engl. J. Med. *311:* 536–537 (1984).

18 Purtilo, D.; Sakamoto, K.; Barnabei, V.; Seeley, J.; Bechtold, T.; Rogers, G.; Yetz, J.; Harada, S.: Epstein-Barr virus-induced diseases in boys with the X-linked lymphoproliferative syndrome (XLP). Update on studies of the registry. Am. J. Med. *73:* 49–56 (1982).

19 Ho, M.; Miller, G.; Atchison, W.; Breinig, M.; Dummer, J.; Andiman, W.; Starzl, T.; Eastman, R.; Griffith, B.; Hardesty, R.; Bahnson, H.; Hakala, T.; Rosenthal, J.: Epstein-Barr virus infections and DNA hybridization studies in posttransplantation lymphoma and lymphoproliferative lesions: the role of primary infection. J. Infect. Dis. *152:* 876–886 (1985).

20 Chaganti, R.; Jhanwar, S.; Koziner, B.; Arlin, Z.; Mertelsmann, R.; Clarkson, B.: Specific translocations characterize Burkitt's-like lymphoma of homosexual men with the acquired immunodeficiency syndrome. Blood *61:* 1269–1272 (1983).

21 Magrath, I.; Erikson, J.; Whang-Peng, J.; Sieverts, H.; Armstrong, G.; Benjamin, D.; Triche, T.; Alabaster, O.; Croce, C.: Synthesis of kappa light chains by cell lines containing an 8; 22 chromosomal translocation derived from a male homosexual with Burkitt's lymphoma. Science *222:* 1094–1098 (1983).

22 Petersen, J.; Tubbs, R.; Savage, R.; Calabrese, L.; Proffitt, M.; Manolova, Y.; Manolov, G.; Shumaker, A.; Tatsumi, E.; McClain, K.; Purtilo, D.: Small noncleaved B cell Burkitt-like lymphoma with chromosome t(8; 14) translocation and Epstein-Barr virus nuclear-associated antigen in a homosexual man with acquired immune deficiency syndrome. Am. J. Med. *78:* 141–148 (1985).

23 Ernberg, I.; Bjorkholm, M.; Zech, L.; Sandstedt, B.; Szigeti, R.; Andersson, J.; Henle, W.; Klein, G.: An EBV genome carrying pre-B cell leukemia in a homosexual man with characteristic karyotype and impaired EBV-specific immunity. J. Clin. Oncol. *4:* 1481–1488 (1986).

24 Bernheim, A.; Berger, R.: Cytogenetic studies of Burkitt lymphoma-leukemia in patients with acquired immunodeficiency syndrome. Cancer Genet. Cytogenet. *32:* 67–74 (1988).

25 Subar, M.; Neri, A.; Inghirami, G.; Knowles, D.; Dalla-Favera, R.: Frequent c-*myc* oncogene activation and infrequent presence of Epstein-Barr virus genome in AIDS-associated lymphoma. Blood *72:* 667–671 (1988).

26 Ziegler, J.; Drew, W.; Miner, R.; Mintz, L.; Rosenbaum, R.; Gershow, J.; Lennette, E.; Greenspan, J.; Shillitoe, E.; Beckstead, J.; Casavant, C.; Yamamoto, K.: Outbreak of Burkitt's-like lymphoma in homosexual men. Lancet *ii:* 631–633 (1982).

27 Groopman, J.; Sullivan, J.; Mulder, C.; Ginsburg, D.; Orkin, S.; O'Hara, C.; Falchuk, K.; Wong-Staal, F.; Gallo, R.: Pathogenesis of B cell lymphoma in a patient with AIDS. Blood *67:* 612–615 (1986).

28 Lind, S.; Gross, P.; Andiman, W.; Stone, S.; Schooley, R.; Harris, N.: Malignant lymphoma presenting as Kaposi's sarcoma in a homosexual man with the acquired immunodeficiency syndrome. Ann. Intern. Med. *102:* 338–340 (1985).

29 Ernberg, I.; Altiok, E.: The role of Epstein-Barr virus in lymphomas of HIV-carriers. APMIS, suppl. *8:* 58–61 (1989).

30 Stansfeld, A.; Diebold, J.; Noel, H.; Kapanci, Y.; Rilke, F.; Kelenyi, G.; Sundstrom, C.; Lennert, K.; van Unnik, J.; Mioduszewska, O.; Wright, D.: Updated Kiel Classification for lymphomas. Lancet *i:* 292–293 (1988).

31 Polack, A.; Hartl, G.; Zimber, U.; Freese, U.; Laux, G.; Takaki, K.; Hohn, B.; Gissman, L.; Bornkamm, G.: A complete set of overlapping cosmid clones of M-ABA

virus derived from nasopharyngeal carcinoma and its similarity to other Epstein-Barr virus isolates. Gene *27:* 279–288 (1984).

32 Rueger, R.; Bornkamm, G.; Fleckenstein, B.: Human cytomegalovirus DNA sequences with homologies to the cellular genome. J. Gen. Virol. *65:* 1351–1364 (1984).

33 Rigby, P.; Dieckmann, M.; Rhodes, C.; Berg, P.: Labelling deoxyribonucleic acid to high specific activity in vitro by nick translation with DNA polymerase I. J. Mol. Biol. *113:* 237–251 (1977).

34 Brigati, D.; Myerson, D.; Leary, J.; Spalholz, B.; Travis, S.; Fong, C.; Hsiung, G.; Ward, D.: Detection of viral genomes in cultured cells and paraffin-embedded tissue sections using biotin-labeled hybridization probes. Virology *126:* 32–50 (1983).

35 Niedobitek, G.; Hamilton-Dutoit, S.; Herbst, H.; Finn, T.; Vetner, M.; Pallesen, G.; Stein, H.: Identification of Epstein-Barr virus infected cells in tonsils of acute infectious mononucleosis by in situ hybridization. Human Pathol. *20:* 796–799 (1989).

36 Rentrop, M.; Knapp, B.; Winter, H.; Schweizer, J.: Aminoalkylsilane-treated glass slides as support for in situ hybridization of keratin cDNAs to frozen tissue sections under varying fixation and pretreatment conditions. Histochem. J. *18:* 271–276 (1986).

37 Frizzera, G.; Hanto, D.; Gajl-Peczalska, K.; Rosa, J.; McKenna, R.; Sibley, R.; Holahan, K.; Lindquist, L.: Polymorphic diffuse B-cell hyperplasias and lymphomas in renal transplant recipients. Cancer Res. *41:* 4262–4279 (1981).

38 Greenspan, J.; Greenspan, D.; Lennette, E.; Abrams, D.; Conant, M.; Petersen, V.; Freese, U.: Replication of Epstein-Barr virus within the epithelial cells of oral 'hairy' leukoplakia, an AIDS-associated lesion. N. Engl. J. Med. *313:* 1564–1571 (1985).

39 Nalesnik, M.; Jaffe, R.; Starzl, T.; Demetris, A.; Porter, K.; Burnham, J.; Makowka, L.; Ho, M.; Locker, J.: The pathology of posttransplant lymphoproliferative disorders occurring in the setting of cyclosporine A-prednisone immunosuppression. Am. J. Pathol. *133:* 173–192 (1988).

40 Spector, B.; Perry, G.; Kersey, J.: Genetically determined immunodeficiency diseases (GDID) and malignancy: report from the immunodeficiency-cancer registry. Clin. Immunol. Immunophatol. *11:* 12–29 (1978).

41 Schneck, S.; Penn, I.: De-novo brain tumours in renal transplant recipients. Lancet *i:* 983–986 (1971).

42 Burkitt, D.; O'Conor, G.: Malignant lymphoma in African children. I. A clinical syndrome. Cancer *14:* 258–269 (1961.

43 List, A.; Greco, F.; Vogler, L.: Lymphoproliferative diseases in immunocompromised hosts: the role of Epstein-Barr virus. J. Clin. Oncol. *5:* 1673–1689 (1987).

44 Purtilo, D.: Epstein-Barr virus-induced onocogenesis in immune-deficient individuals. Lancet *i:* 300–303 (1980).

45 Klein, G.: Lymphoma development in mice and humans: diversity of initiation is followed by convergent cytogenetic evolution. Proc. Natl. Acad. Sci. USA *76:* 2442–2446 (1979).

46 Nilsson, K.; Klein, G.: Phenotypic and cytogenetic characteristics of human B-lymphoid cell lines and their relevance for the etiology of Burkitt's lymphoma. Adv. Cancer Res. *37:* 319–387 (1982).

47 Hochberg, F.; Miller, G.; Schooley, R.; Hirsch, M.; Feorino, P.; Henle, W.: Central-nervous-system lymphoma related to Epstein-Barr virus. N. Engl. J. Med. *309:* 745–748 (1983).

48 Pulford, K.; Falini, B.; Heryet, A.; Gatter, K.; Mason, D.: 4KB5, a new monoclonal

anti-B-cell antibody for the routine diagnosis of lymphoid tissue biopsies; in McMichael (ed): Leucocyte Typing III, p. 828 (Oxford University Press, Oxford 1987).
49 Epstein, A.; Marder, R.; Winter, J.; Fox, R.: Two new monoclonal antibodies (LN-1, LN-2) reactive with B5 formalin fixed, paraffin embedded tissues with follicular centre and mantle zone human B lymphocytes and derived tumours. J. Immunol. *133:* 1028–1036 (1984).
50 Ishii, Y.; Takami, T.; Takei, T.; Kikuchi, K.: Two distinct antigen systems in human B lymphocytes: identification of cell surface and intracellular antigens using monoclonal antibodies. Clin. Exp. Immunol. *58:* 183–192 (1984).
51 Smith, S.; Brown, M.; Rowe, D.; Callard, R.; Beverley, P.: Functional subsets of human helper inducer cells defined by a new monoclonal antibody UCHL1. Immunology *58:* 63–70 (1986).
52 Poppema, S.; Holema, H.; Visser, L.; Vos, H.: Monoclonal antibodies (MT1, MT2, MB1, MB2, MB3) reactive with leukocyte subsets in paraffin embedded tissue sections. Am. J. Pathol. *127:* 418–429 (1987).

Dr. S. J. Hamilton-Dutoit, MRCPath, Laboratory of Immunohistology, University Institute of Pathology, Kommunehospitalet, Finsensgade 12, DK–8000 Aarhus C (Denmark)

Racz P, Haase AT, Gluckman JC (eds): Modern Pathology of AIDS and Other Retroviral Infections. Basel, Karger, 1990, pp 130–141

Polymerase Chain Reaction: A Powerful Diagnostic Tool

Vijay A. Varma, David Swan

Emory University and Centers for Disease Control, Atlanta, Ga., USA

Rapid advances in molecular biology during the past decade have led to detailed understanding of the genomic structure of many infectious agents including pathogenic viruses. This information has been used to generate probes for hybridization assays for the diagnosis of infectious diseases. These assays have supplemented viral cultures and serologic assays for diagnostic and epidemiology studies. However, all these methods have limited sensitivity particularly for viral diseases with low copy number such as human immunodeficiency virus (HIV).

Polymerase chain reaction (PCR), a novel method to synthesize multiple copies of DNA in vitro [17], has proved to be an extremely sensitive method for detection of viruses and other pathogens in clinical samples. The DNA of interest is first amplified by this method to a level that can be readily identified by hybridization. Thus even a single copy of viral genome in a patient's sample can theoretically be amplified a millionfold and then identified with high specificity. This technique has proven valuable in the detection of HIV, a virus that can be latent for prolonged periods and may be sequestered in a few cells.

Here we describe briefly the PCR technique, review its applications and, using HIV disease as an example, discuss its value as a diagnostic tool.

Principle

PCR is based on a simple three-step process: (1) denaturation of double-stranded DNA to single strands; (2) annealing of primers to these separated strands, and (3) extension of these primers resulting in a copy of each of the DNA strands.

These three steps constitute a cycle and with each repetition of the cycle the targeted DNA is doubled resulting in exponential amplification (fig. 1).

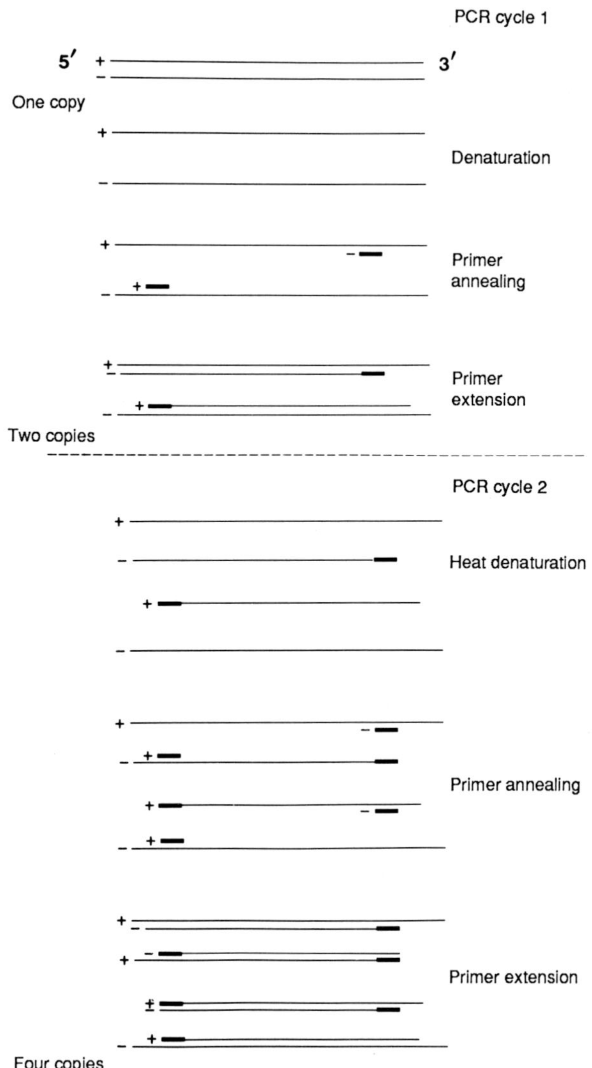

Fig. 1. Principle of enzymatic amplification of DNA, in vitro, by PCR. The first two cycles resulting in fourfold amplification of targeted sequence are shown.

Thus if this cycle is repeated 20 times, theoretically a million copies of the DNA can be made and with 30 cycles a billion. Even though the yield is less than these theoretical numbers in practice, a 30,000-fold amplification of a DNA segment can be readily achieved within a few hours.

Preparation of Sample

DNA from plasmid and phage vectors, viral cultures, blood cells and paraffin-embedded tissue have all been successfully amplified by PCR. With this technique, DNA can be amplified starting with a very minute sample; there is virtually no lower limit to sample size. Specific sequences have been amplified starting with as little as a single 5-μm section of a $2 \times 2 \times 2\ mm^3$ tissue biopsy, a single hair, a single cell and a single sperm which has only half the normal diploid cellular DNA [12].

Another distinct advantage of this technique is that DNA need not be extracted in a pure form [17]. A crude lysate will suffice and the DNA of interest may be degraded and may constitute a small fraction of a crude whole tissue lysate. The primers do not appear to have any problem in finding the needle (target sequence) in the hay stack (heterogeneous mix of other DNA, proteins and lipids) and initiating the chain reaction; these extraneous and cellular components do not appear to interfere with the enzymatic activity of Taq.

DNA purified by the classical phenol-chloroform extraction method can be used for PCR but such purity is unnecessary particularly for diagnostic work. This method is costly and laborious. Furthermore, during the manipulations required for the extraction of DNA, substantial fractions of the DNA may be lost and this may be critical in small clinical samples. We have successfully used the following three methods for preparation of various biologic samples for PCR:

(1) *Boiling:* Blood, serum or deparaffinized tissue is boiled in a small aliquot of distilled water and the supernatant is used for PCR reaction. For small samples it is best to boil the specimen and add other PCR components and carry out the reaction in the same microcentrifuge tube. Boiling appears to be sufficient to lyse the cells and bring at least a portion of the DNA into solution.

(2) *Protease digestion:* The sample is digested in proteinase K (500 μg/ml) in phosphate-buffered saline (PBS) overnight at 37 °C. This mixture is then boiled to inactivate the protease.

(3) *Nonionic detergent and protease digestion* [9]: Incubate the specimen in a mixture of nonionic detergents (0.45% Nonidet P40, 0.45% Tween 20,

0.1 mg/ml gelatin, 2.5 m*M* $MgCl_2$, 10 m*M* Tris-HCl (pH 8.3) and 50 m*M* KCl) and proteinase K 0.6 µl of 10 mg/ml proteinase K at 50–60 °C for 1 h. The protease is inactivated by incubating this mix at 95 °C for 10 min. The nonionic detergent solution can be prepared, autoclaved and stored frozen but the proteinase K must be added fresh just before the incubation.

Though all three procedures work, the last method with nonionic detergents appears to give the best yield from a given sample. For paraffinized tissue, a single 5-µm section is obtained, deparaffinized, washed in ethanol, dried and prepared by one of the above methods. We found that boiling is less efficient and a whole section may be necessary for one PCR run. After digestion in 100 µl of nonionic detergents and proteinase K, an aliquot of as little as 5 µl is sufficient for PCR amplification.

Design of Primers

The primers are short segments of synthetic single-stranded DNA (oligonucleotides) which act as starting points for the copying of each of the DNA strands. The oligonucleotides, usually 20–30 bases in length, are made employing an automated DNA synthesizer. To achieve specific amplification, the primers should be at least 18 bases long.

The DNA sequence of at least part of the gene to be amplified must of course be known before the primers are designed. A region within this gene, usually 100–500 bases long, is selected. The (+) primer has the same sequence as the (+) or the sense strand of the DNA at the 5′ end of the segment to be amplified. The (–) primer has the same sequence as the (–) or the anti-sense strand of the DNA at the 3′ end of the segment to be amplified. The (+) strand will be complimentary to the (–) strand of the template DNA and the (–) strand will be complimentary to the (+) strand of the template. These primers are then extended from their 3′ ends to assemble a complimentary copy of each of the template strands. In the next cycle these newly synthesized daughter strands will also act as templates resulting in exponential amplification of the DNA segment between the two primers.

Reaction Components

The PCR is carried out in a small volume, typically 100 µl, in a microcentrifuge tube. The reaction mix consists of the specimen DNA, the primers, the four nucleotides which are the building blocks of the DNA, a DNA polymerase which assembles these nucleotides in 5′ to 3′ direction, and a buffer containing magnesium which is essential for activity of the enzyme.

A major advance in PCR technology has been the introduction of DNA polymerase from Thermus aquaticus (Taq) [20]. In the original PCR experiments Klenow, a subunit of *Escherichia coli* DNA polymerase I – the Klenow fragment was used. However, this heatlabile enzyme is destroyed during the heat denaturation step and fresh enzyme has to be added for each new PCR cycle. Unlike Klenow, Taq enzyme is stable at 95 °C. This enzyme can therefore be added to the reaction mix at the beginning and it remains active during repeated denaturation, annealing and extension cycles. This has permitted automation of PCR. Simple and reliable computer-controlled thermal cyclers are now available to run PCR. Because of the high fidelity of this enzyme and its high optimum temperature of 72 °C, the specificity and yield of PCR has also improved with Taq.

Composition of PCR Reaction Mix
(From protocol recommended by Perkin Elmer Cetus):

Taq polymerase: 2.5 units
dATP 200 μ*M*
dCTP 200 μ*M*
dGTP 200 μ*M*
TTP 200 μ*M*
DNA from sample
+ primer 20 pmol
– primer 20 pmol
Reaction buffer: 50 m*M* KCl, 10 m*M* Tris-Cl, pH 8.3, 1.5 m*M* $MgCl_2$ and 0.1% gelatin.
Double distilled water to 100 μl.

Comment. We used this reaction mix for amplification of various DNA templates from a variety of clinical samples. However, to optimize the reaction for a specific target in a specific type of sample the concentrations of Mg, Taq and deoxyribonucleosides (dNTPs) may have to be adjusted.

PCR Reaction Parameters
(From protocol recommended by Perkin Elmer Cetus):

Denaturation at 94 °C for 1 min.
Annealing of primers at 50 °C for 2 min.
Extension of primers at 72 °C for 3 min. This cycle is repeated 30–35 times.

Comments. These cycle parameters should suffice for amplification of most target DNA samples. An additional initial denaturation step by heating the sample at 94 °C for 7 min may improve the yield. The annealing temperature may be raised to improve the specificity of the reaction; at higher temperatures the stringency of hybridization of the primer to template is increased and then it will anneal only to the exact complementary sequence. If the primers are small, a slow raising or ramping of the temperature from annealing to extension temperatures is recommended. During this slow ramping the primer begins to extend and at the optimal temperature of 72 °C this extended primer-template hybrid is more thermostable allowing completion of the copy.

The number of cycles should be limited to the minimum number required to achieve the desired amplification. Though, theoretically, the DNA is amplified exponentially with each additional cycle, the efficiency of the reaction is progressively reduced by the accumulation of the amplified sequences competing with primers, and by the depletion of the primers and nucleotides. Furthermore, excessive cycles tend to produce additional nonspecific products.

Detection of Amplified DNA

A variety of methods can be used to detect the amplified DNA. Usually PCR yields an enormous number of copies of the targeted DNA and these can be visualized on an ethidium bromide stained agarose gel. Since almost all these copies are of defined length, a distinct band is readily identified; since the size (number of base pairs) of the amplified DNA is determined by the design of primers, this band will be of the predefined size. To confirm that this amplification has indeed been specific, this DNA can be digested with a restriction enzyme which is known to cut the target at a specific site yielding fragments of known size. Alternatively, a Southern blot can be obtained by transferring the DNA from the gel onto a filter and hybridization with a P-32-labelled probe complementary to a sequence between the two primer sequences can be performed.

Oligomer restriction is a further refinement of this detection method [13, 18]. A synthetic DNA probe is designed such that it encompasses a restriction site within the amplified DNA. It is labelled at its 5′ end with P-32 and hybridized with the amplified DNA in solution. When this DNA is digested with the specific restriction enzyme, it will cleave only the double-stranded DNA including the hybrid of radiolabelled probe and the amplified sequence.

This resulting probe fragment bearing the radioactive label can be readily detected on a Southern blot.

A simple and more rapid detection method is based on dot or slot blot technique [21]. Here the amplified DNA is directly spotted onto the filter and hybridized with an oligonucleotide probe; again the probe is designed to be complementary to a sequence between the two primers. A colorimetric method has also been developed to avoid the use of radioactive reagents [3]. Here the probe is labelled with biotin or digoxigenin and positive hybridization is detected by a color reaction.

Applications

The PCR, because of its elegant simplicity and versatility, has quickly become an essential technique to molecular biologists in a wide variety of settings [16]. It has been used to diagnose a variety of inherited diseases including hemoglobinopathies, phenylketonurea and alpha-1-antitrypsin deficiency [1, 4]. It has been used to detect residual disease in patients treated for lymphoma and it has been used for analysis of ras oncogenes in lung, colon, pancreas and other organs [2, 5]. It has been used for HLA typing and to establish individual identity in forensic settings, and to examine ancient DNA from a wooly mammoth and an Egyptian mummy [16].

In the clinical setting, PCR offers an exquisitely sensitive diagnostic assay for detection of viruses and other pathogens. Compared to direct hybridization methods, including in situ hybridization, this technique provides sensitivity which is greater by several orders of magnitude. Shibata et al. [22] tested the sensitivity of PCR in detecting the human papilloma virus (HPV). Each cell of the SiHa cell line has one or two copies of HPV 16 genome. These were not detectable by in situ hybridization. When paraffin-embedded HiHa cells were assayed by PCR, an average of 0.1–0.2 genomes/cell were detected.

A particularly valuable clinical application has been its use in the diagnosis of AIDS [6]. This disease caused by two retroviruses of the Lentivirinae subfamily, human immunodeficiency virus (HIV) type 1 and type 2. The virus enters T4 lymphocytes and macrophages via CD4 receptors [7]. Inside the cell a DNA copy of the viral RNA genome is made; this DNA integrates into the host genome and this proviral form usually remains dormant for varying periods before the cell is activated leading to HIV replication.

In individuals infected with HIV, the virus is usually present in low numbers, may be sequestered in a few cells, and lie dormant for varying lengths of time. During these periods of latency, the widely used serologic assays for HIV antibodies and antigens may be negative. Cultures may detect the virus at this stage of the disease, but retroviral culture methods are lengthy and cumbersome and expensive. PCR offers a rapid and reliable tool for detection of the virus in these patients during this serologic window between infection and seroconversion: the virus has been identified in some individuals as early as 35 months before seroconversion [10].

PCR has also proven valuable in the diagnosis of HIV infection in the pediatric population [15]. Up to 50% of babies born to HIV-infected mothers acquire this virus in utero, during birth or later through the maternal milk. The signs and symptoms of HIV disease in these babies are nonspecific. However, the main diagnostic problem is the limited reliability of serologic assays in this group. Whether infected by the virus or not, these babies may have high titers of HIV antibodies acquired from the mother in utero; these antibodies may last for as long as 15 months. Because of the excess of these antibodies the assays for the HIV antigen are also less sensitive. PCR has been used to diagnose HIV infection in this high risk infant group by direct identification of the viral sequences [19].

Though PCR amplifies DNA it can also be used to detect specific RNA sequences [8]. A complementary copy (cDNA) of the RNA is first made using a reverse transcriptase enzyme. These cDNA molecules are then amplified using the same three-step PCR cycles described before. Assays for HIV RNA by this method may prove valuable to study the viral transcriptional activity during various phases of the disease.

As noted before, PCR can be used to amplify DNA from paraffin-embedded tissue samples [11]. Though the DNA is partially degraded in these samples, HIV-specific sequences have been successfully amplified from paraffin-embedded autopsy and surgical biopsy tissue samples [14, 24, 26]. Except for the initial deparaffinization step, all subsequent steps, i.e., preparation of the sample, reaction mix, reaction parameters and detection of amplified DNA are similar to those described before (fig. 2). Paraffin-embedded tissue samples have also been used to detect other viruses, particularly HPV [21–23].

Nucleic acid analysis of paraffin-embedded tissue employing PCR has immense potential [23]. There are vast archives of well-documented paraffin-embedded tissue samples worldwide, and these could be used for detection and study of various pathogens that are otherwise difficult to detect. In the case of AIDS for example, the only material available on most of the cases from the early period of the epidemic are paraffin blocks archived in pathology departments. The majority of these cases were diagnosed on clinical criteria

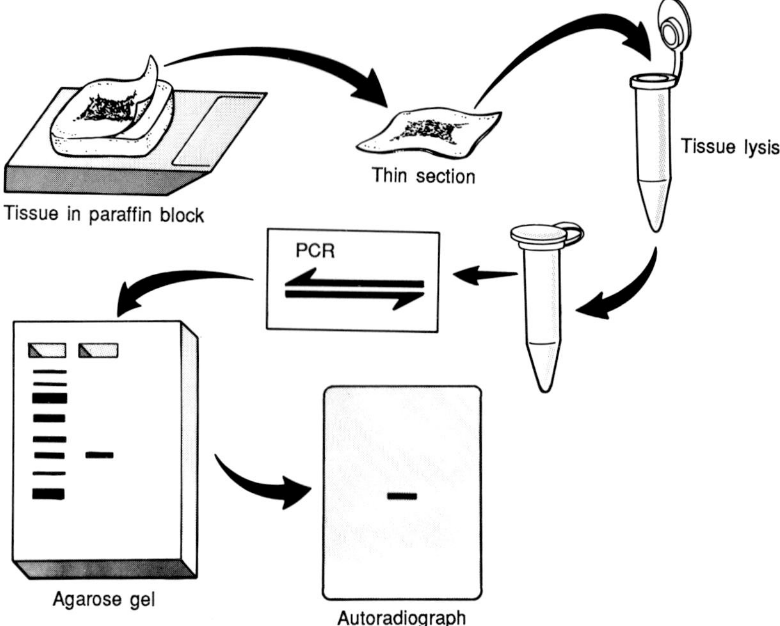

Fig. 2. PCR on paraffin-embedded tissue. Preparation of the sample, amplification of specific sequence, identification of this amplified product on an agarose gel, and confirmation of the specificity of amplification on a Southern blot by autoradiogram are depicted.

and the diagnosis was never confirmed. The virus can now be detected and the diagnosis established employing PCR on these paraffin blocks.

We have used PCR to amplify DNA from material recovered from stained sections on archived glass slides [25]. The coverslips are removed, the sections are destained, and then scraped with a clean scalpel blade into a microcentrifuge tube. This material is digested using one of the methods described before and subjected to PCR. Using this technique we amplified both viral and genomic sequences from a variety of histologic slides stored for as long as 40 years.

PCR, however, is more than a sensitive detection system. The amplified sequences can be used for detailed molecular analysis. Two potential applications are: (1) DNA sequencing of PCR-amplified segments to analyze the different strains within different tissues of the same patient, to compare the strains from population groups from different periods of an epidemic and from different geographic areas. (2) Recombinant experiments employing genetically important segments of the viral sequence such as promoter regions, recovered from clinical samples by PCR, to understand the molecular basis of tissue tropism and virulence of different viral strains.

Precaution

PCR is a simple yet powerful technique that has the potential of wide clinical use. However, its main strength, the capability of generating millions of copies from a single DNA molecule, is also its main pitfall. Most workers have experienced false-positive results due to contamination problems. Even a minute amount (5 pg or less) of the DNA of interest, contaminating another negative sample, will be amplified leading to erroneously positive results. The chief source of contamination is PCR products from previous experiments. If by accident these are carried over into the sample or any of the reaction mix components in another experiment, they will be efficiently amplified and the product will be indistinguishable from that amplified from the sample DNA. It is therefore prudent to provide physically separate areas for preparation of the sample, running the PCR and handling the PCR product. Separate sets of positive displacement pipettes, glassware and disposables should be used in these three areas.

Other sources of contaminations include, DNA from cultures, plasmids, and other specimens. Between each paraffin block used for PCR, the microtome blade should be thoroughly cleaned to prevent carry-over from one specimen to another.

In summary: PCR, a recently invented technique for amplification of specific segments of DNA, has been adapted for detection of various infectious agents. It is a fairly simple and rapid assay which offers high sensitivity and specificity. With the introduction of automation and colorimetric methods for detection of amplified product, it may become more amenable to clinical laboratories. This method is particularly useful for detection of pathogens that are in low copy number such as HIV. Because of its ability to amplify specific sequences from DNA in paraffin-embedded tissue, this method offers a way to conduct detailed retrospective studies involving analysis of genomic DNA or the nucleic acids of infectious agents, in archived pathology material. In addition to being a powerful diagnostic detection system, it is an invaluable research tool for studies in molecular biology of inherited, neoplastic and infectious diseases.

References

1 Abbott, CM, McMahon CJ, Whitehouse DB, et al: Prenatal diagnosis of alpha-1-antitrypsin deficiency using polymerase chain reaction. Lancet 1988;i:763–764.
2 Almoguera C, Shibata D, Forrestor K, et al: Most human carcinomas of the exocrine pancreas contain mutant c-K-ras genes. Cell 1988;53:549–554.
3 Bugawan TL, Saiki RK, Levenson CH, et al: The use of non-radioactive oligonucleotide probes to analyze enzymatically amplified DNA for prenatal diagnosis and forensic HLA typing. Biotechnology 1988;6:943–947.
4 Dilella AG, Huang WM, Woo SLC: Screening for phenylketonuria mutations by DNA amplification with the polymerase chain reaction. Lancet 1988;i:497–499.
5 Forrestor K, Almoguera C, Han K, et al: Detection of high incidence of K ras oncogenes during human colon tumorigenesis. Nature 1987;327:298–303.
6 Guatelli JC, Gingeras TR, Richman DD: Nucleic acid amplification in vitro: Detection of sequences with low copy numbers and application to diagnosis of human immunodeficiency virus type 1 infection. Clin Microbiol Rev 1989;2:217–226.
7 Harper MH, Marselle LM, Gallo RC, et al: Detection of lymphocytes expressing human T-lymphotropic virus type III in lymph nodes and peripheral blood from infected individuals by in situ hybridization. Proc Natl Acad Sci USA 1986;83:772–776.
8 Hart C, Schochetman G, Spira T, et al: Direct detection of HIV RNA expression in seropositive persons. Lancet 1988;ii:596–599.
9 Higuchi R: Rapid, efficient DNA extraction for PCR from cells or blood. Amplifications 1989;2:1–3.
10 Imagawa DT, Lee MH, Wolinsky SM, et al: Human immunodeficiency virus type 1 infection in homosexual men who remain seronegative for prolonged periods. N Engl J Med 1989;320:1458–1462.
11 Impraim CC, Saiki RK, Erlich HA, et al: Analysis of DNA extracted from formalin-fixed, paraffin-embedded tissues with enzymatic amplification and hybridization with sequence-specific oligonucleotides. Biochem Biophys Res Commun 1987;142:710–716.
12 Kumar R, Barbacid M: Oncogene detection at the single cell level. Oncogene 1988;3:647–651.
13 Kwok S, Mack DH, Mullis KB, et al: Identification of human immunodeficiency virus sequences by using in vitro enzymatic and oligomer cleavage detection. J Virol 1987;61:1690–1694.
14 Lai-Goldman M, Lai E, Grody WW: Detection of human immunodeficiency virus (HIV) infection in formalin-fixed, paraffin-embedded tissues by DNA amplification. Nucl Acids Res 1988;16:8191.
15 Laure F, Courgnaud V, Rouzioux C, et al: Detection of HIV I DNA in infants and children by means of the polymerase chain reaction. Lancet 1988;ii:538–541.
16 Marx JL: Multiplying genes by leaps and bounds. Science 1988;240:1408–1410.
17 Mullis KB, Faloona FA: Specific synthesis of DNA in vitro via a polymerase-catalyzed chain reaction. Methods Enzymol 1987;155:335–350.
18 Ou C, Kwok S, Mitchell SW, et al: DNA amplification for direct detection of HIV-1 in DNA of peripheral blood mononuclear cells. Science 1988;239:295–297.
19 Rogers MF, Ou C-F, Rayfield M, et al: Use of the polymerase chain reaction for early detection of the proviral sequences of human immunodeficiency virus in infants born to seropositive mothers. N Engl J Med 1989;320:1649–1654.

20 Saiki et al: Primer-directed enzymatic amplification of DNA with thermostable DNA polymerase. Science 1988;239:487–491.
21 Shibata DK, Arnheim N, Martin JW: Detection of human papilloma virus in paraffin-embedded tissue using the polymerase chain reaction. J Exp Med 1988;167:225–230.
22 Shibata D, Fu YS, Gupta JW, et al: Detection of human papilloma virus in normal and dysplastic tissue by the polymerase chain reaction. Lab Invest 1988;59:555–559.
23 Shibata D, Martin WJ, Arnheim N: Analysis of DNA sequences in forty-year-old paraffin-embedded thin tissue sections: A bridge between molecular biology and classical histology. Cancer Res 1988;48:4564–4566.
24 Varma VA, Hunter S, Tickman R, et al: Acute fatal encephalitis with negative serologic assays for antibody and antigen: Diagnosis by polymerase chain reaction. NEJM 1989;320:1494–1495.
25 Varma VA, Tickman R, Bain R, et al: Enzymatic amplification of specific human genomic sequences: polymerase chain reaction in formalin-fixed, paraffin-embedded, hemotaxylin-eosin-stained sections. Lab Invest 1989;60:100A.
26 Varma VA, Rimland D, Srinivasan A, et al: Diagnosis of HIV infection employing polymerase chain reaction (PCR) on paraffin-embedded tissue. Lab Invest 1989;60:101A.

Vijay A. Varma, MD, Department of Pathology, Lab (113), VA Medical Center, 1670 Clairmont Road, Atlanta, GA 30033 (USA)

Racz P, Haase AT, Gluckman JC (eds): Modern Pathology of AIDS and Other Retroviral Infections. Basel, Karger, 1990, pp 142–154

Use of in situ Hybridization for Studies of AIDS Pathogenesis in the SIV System

M. S. Wyand[a], *R. C. Desrosiers*[b], *N. W. King*[b]

[a]EG&G Mason Research Institute, Worcester, Mass; [b]Harvard Medical School, New England Regional Primate Research Center, Southborough, Mass., USA

The epidemic of human acquired immunodeficiency syndrome (AIDS) has focused considerable attention on the pathogenesis of retroviral diseases. Several features of lentiviruses, including HIV, such as the long period of persistent infection and the resistance of lentiviral diseases to traditional vaccines, reemphasize the need for meaningful animal model systems. Although HIV has been inoculated into a number of species such as chimpanzees [1, 14–16], rabbits [12], and the SCID-hu mouse [25], none of these attempts has resulted in the development of disease in the recipient species analogous to AIDS in humans. The simian immunodeficiency viruses (SIVs) are closely related to the HIVs and some strains induce a fatal disease in rhesus monkeys very similar to AIDS. SIV infection of rhesus monkeys has thus emerged as one of the most important AIDS animal model systems.

SIV was first isolated in 1985 at the New England Regional Primate Research Center from rhesus monkeys that had been inoculated with cell-free lymphoma homogenates as part of tumor transmission experiments [9]. In addition to developing lymphoma, some animals developed opportunistic infections such as cryptosporidiosis, and disseminated cytomegalovirus infection. Similar viruses were isolated at other primate centers from immunodeficient rhesus monkeys inoculated with homogenates of cutaneous lepromatous leprosy from a sooty mangabey monkey [2, 24], from healthy sooty mangabeys [13, 22], and from pigtailed macaques [3]. The isolates are now designated as SIV followed by letters denoting the species from which they were isolated (SIVmac the original isolate from rhesus macaques, SIV/SMM from sooty managabeys) or the institution at which they were isolated (SIV/Delta, the Delta Primate Center). Recent information indicates that SIV is not found in rhesus monkeys in the wild and that it does not cause disease in the species which naturally harbor the virus such as mangabeys and green monkeys.

Like HIV, SIV is a lentivirus 110–130 nm in diameter with a cylindrical nucleoid. The virus replicates preferentially in CD4-positive cells and can be isolated with HUT-78 cells, H9 cells, or IL-2-stimulated human or monkey PBLs [9, 18]. SIV proteins of 160, 120, 55 and 24 kd, similar in size to the major gag and env proteins of HIV, can be identified by radioimmunoprecipitation [17]. Monoclonal antibodies to the p24 major core protein of HIV will recognize the p27 major core protein of SIV [17, 26] and serum from SIV-infected monkeys and HIV-infected humans will cross-react with the gag precursor, p55. SIV-infected rhesus monkeys develop a progressive fatal immunosuppressive disease [21]. Animals that die early in the course of infection develop low titer anti-SIV antibodies, have low plasma IgG levels, and low absolute numbers of CD4-positive peripheral blood lymphocytes [19]. In contrast, animals that become persistently infected for prolonged periods [10] develop high titer antibody responses which recognize both core and envelope proteins, and have elevated plasma IgG and delayed decreases in the number of circulating CD4 cells. Early in the course of infection, monkeys develop a peripheral lymphadenopathy similar to the prodromal syndrome of progressive generalized lymphadenopathy in humans [7]. Later, as the course of immunodeficiency progresses, hyperplastic germinal centers involute, the number of CD8-positive cells in the paracortex becomes greater than the number of CD4-positive cells, and progressive lymphoid depletion occurs [8]. Proliferative responses of PBLs to mitogens are also decreased [21]. Monkeys may also develop an erythematous skin rash similar to that seen in humans [28, 29].

Rhesus monkeys experimentally infected with SIV develop lesions similar to those seen in human AIDS patients. Monkeys have severe weight loss and wasting with thymic atrophy, lymphoid depletion and a characteristic granulomatous meningoencephalitis [21, 30]. Opportunistic infections include pneumocystis pneumonia, disseminated cryptosporidosis, cytomegalovirus infection, disseminated *Mycobacterium avium intracellulare* infection, adenoviral pancreatitis, and cutaneous and oral candidiasis. Other lesions for which etiologic agents have not been identified are subacute portal hepatitis, membranoproliferative glomerulonephritis, chronic gastritis, lymphocytic enteritis with villous atrophy, and chronic colitis with crypt abscesses.

We have used the SIV-infected rhesus monkey to study the molecular biology and pathogenesis of retroviral-induced immunodeficiency. These studies have included an examination of the pathology of SIV-associated disease and the use of in situ hybridization to localize SIV RNA in brain and lymphoid tissues. Brain and lymphoid tissues undergo pathologic changes in human or simian AIDS which are thought to be directly related to virus infection and not to opportunistic pathogens. For our studies we developed sensitive in situ hybridization methods using formalin-fixed tissues and DNA probes, since formalin-fixed tissues are most easily obtained at necropsy and DNA probes have greater stability than RNA probes.

A SIVmac subclone containing envelope sequences was nick translated and used for tissue hybridizations. The subclone, pENVSS142 (Bluescript vector, Stratagene) contained a 3.48 kilobase Sstl-Sstl SIV envelope sequence, nucleotide sequence 5241–8718 [6]. Plasmid pBR322 and lambda phage were used without inserts as negative controls. Subclones and plasmid controls were labeled by nick translation using procedures modified from Maniatis et al. [23] and Rigby et al. [27]. Tritium-labelled deoxythymidine and deoxycytidine were incorporated into the subclone under conditions which produced probes approximately 300–600 base pairs in length with a specific activity of 5×10^7 cpm/μg. We have compared probes labelled with tritium and ^{35}S for in situ hybridization and have found ^{35}S to produce high background signal and to bind nonspecifically to eosinophils in some tissues.

Tissues were collected from rhesus monkeys inoculated with SIVmac that was prepared on either HUT-78 cells or on IL-2-stimulated human peripheral blood lymphocytes [10]. Tissues were collected at necropsy or by biopsy and were fixed in formalin. Formalin-fixed tissues in our laboratory have provided greater signal intensity and better morphology for in situ hybridization than have frozen tissues. After fixation, tissues were embedded in paraffin and were routinely sectioned at 5 μm. The hybridization procedure was a modification of those published by Singer et al. [33] and Brigati et al. [5] – see also Wyand et al. [38]. All hybridizations were performed under nondenaturing conditions to localize SIV RNA. Slides were coated in NTB-2 emulsion and were developed in the dark at 4°C for 2–6 weeks.

Approximately 50% of the rhesus monkeys inoculated with SIVmac at the New England Regional Primate Research Center develop a meningoencephalitis [20, 21, 30]. The most prominent lesions are discrete aggregates of mononuclear cells with round to oval nuclei and variable amounts of eosinophilic cytoplasm. Some of the nodules contain karyorrhexic cellular debris, while in others the mononuclear cells are quite large with abundant pale granular eosinophilic cytoplasm. Giant cells are frequently present within the nodules as well as in the meninges and free within the brain parenchyma. The white matter, especially the deeper regions of the internal capsule, corpus

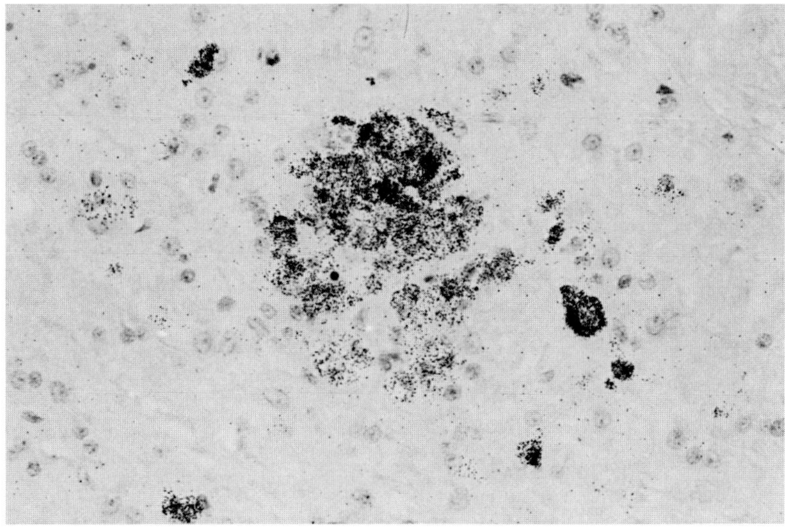

Fig. 1. A SIV envelope subclone pENVSS142 labelled with tritium was used for in situ hybridization to localize SIV RNA under nondenaturing conditions. A focal aggregate of mononuclear cells in the gray matter demonstrates the large amount of SIV RNA typically seen in these lesions. A small vessel bisects this nodule emphasizing the perivascular nature of these aggregates. The exposure for this hybridization was 6 weeks.

callosum, and brain stem, contain more nodules than the cortical gray matter, however the basal ganglia and thalamus are also affected. The nodules are usually perivascular, but in many cases no association with vessels is visible.

In situ hybridization with the SIV DNA probe identified abundant SIV RNA in approximately 60–70% of the mononucleate and multinucleate cells that were present in these lesions (fig. 1). Hybridization signal was present over cells distant from the nodules as well (fig. 2), in areas where no morphologic changes were visible. These cells often had round to angular nuclei with distinct nucleoli and resembled astrocytes or oligodendrocytes. Additional studies are currently in progress to specifically phenotype the cells that contain the abundant RNA in these lesions.

We have also used in situ hybridization to detect SIV RNA in lymphoid tissues collected at necropsy and by biopsy. Rhesus monkeys develop a generalized lymphadenopathy during the early stages of SIV infection similar to the lymphadenopathy seen in human AIDS patients. Peripheral lymph nodes may

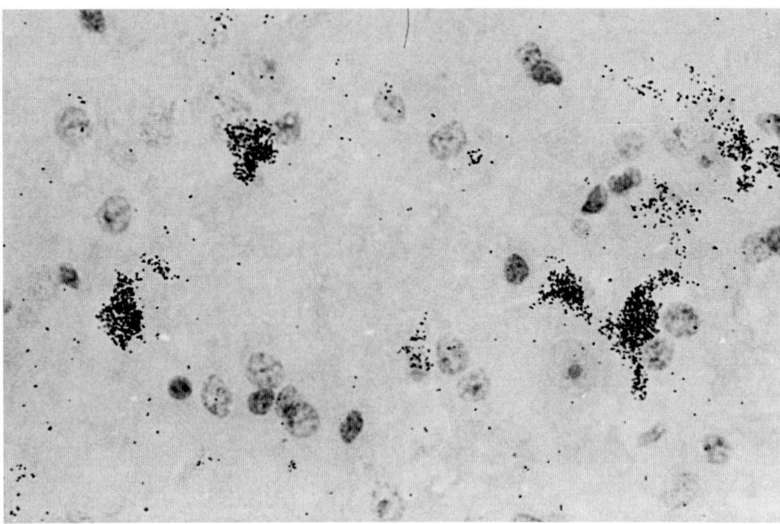

Fig. 2. Individual cells are often present away from mononuclear cell aggregates in areas that have no visible lesion or mild gliosis. These cells frequently have a nuclear morphology compatible with glial cells although their positive identification can be difficult by in situ hybridization alone.

enlarge to more than 3 cm in diameter and contain large irregular hyperplastic follicles with discontinuous mantle zones (fig. 3). Lymph nodes from monkeys were collected for in situ hybridization during stages of follicular hyperplasia by biopsy. Cells containing SIV RNA identified by in situ hybridization were infrequently seen during stages of follicular hyperplasia and were more commonly observed in the interfollicular paracortex and the mantle zones (fig. 4) than within the germinal center. The majority of positive cells were identified as small to medium-sized lymphocytes based upon nuclear morphology and the scant amount of cytoplasm. Occasional positive cells in the paracortex had large vesicular nuclei with prominent nucleoli and a moderate amount of eosinophilic cytoplasm and were most compatible with macrophages.

At necropsy, lymph nodes did not display evidence of follicular hyperplasia but were in various stages of follicular involution and paracortical depletion (fig. 5). In these animals a dramatic number of SIV-positive cells were

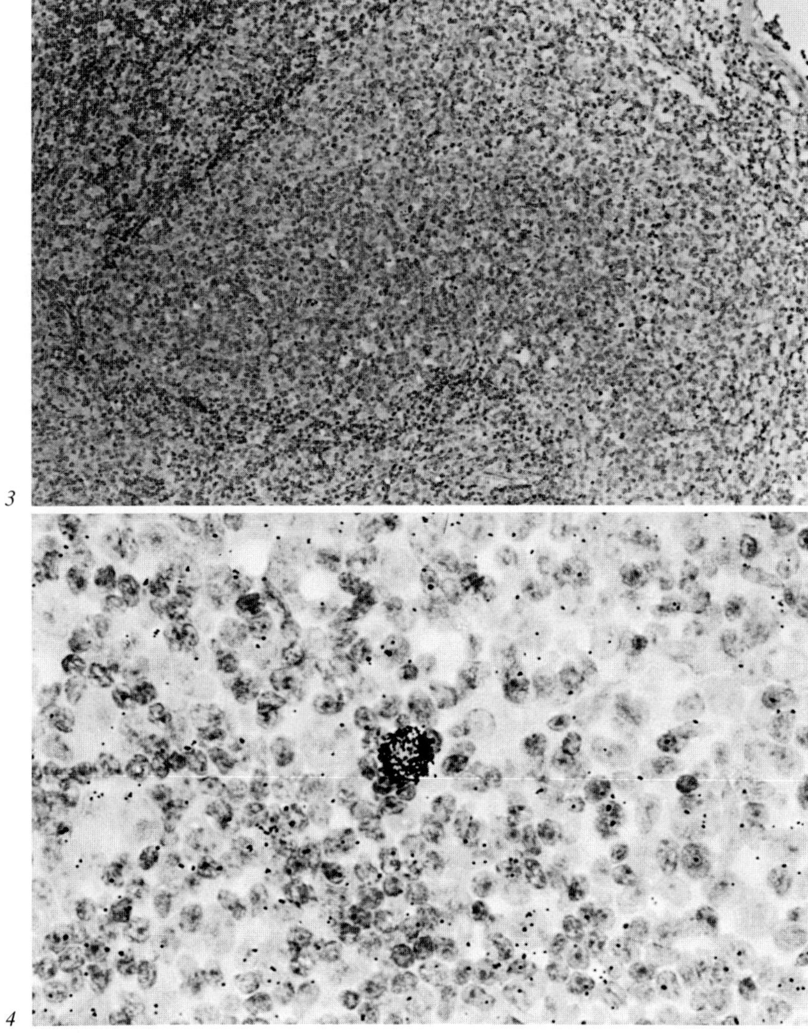

Fig. 3. Irregular hyperplastic follicles expand the lymph node parenchyma during stages of lymphadenopathy. Follicles extend deep within the medulla of the node, often have discontinuous mantle zones and contain hemorrhage. In these stages the number of cells containing SIV RNA detected by in situ hybridization is low and are found mostly in the paracortex or mantle zones. Scattered follicles occasionally contain cells positive for SIV RNA.

Fig. 4. A large cell present in the mantle zone of a hyperplastic follicle with a poorly defined margin. The cell has a central area of decreased grains which represents the nucleus. Abundant silver grains indicate a large amount of SIV RNA in the cytoplasm. The size of the cell and the amount of cytoplasm are compatible with a macrophage.

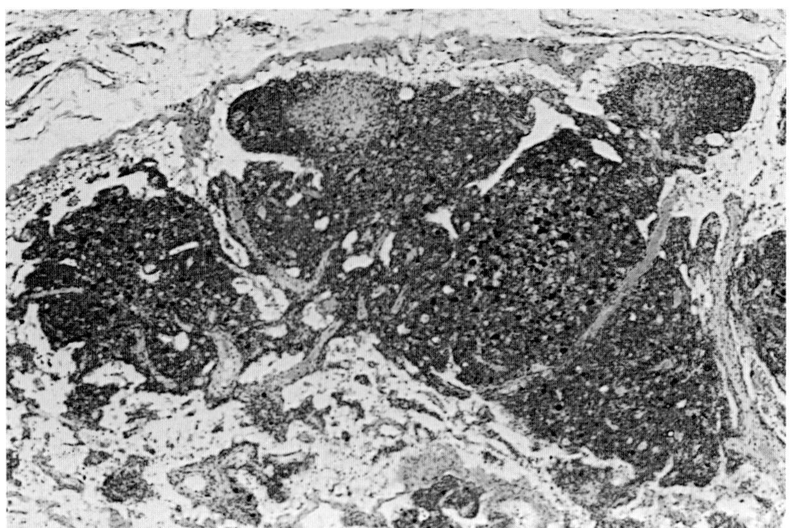

Fig. 5. At necropsy, follicles are either absent or present as hyalinized remnants in the cortex adjacent to the subcapsular sinus. The lymph node parenchyma is often reduced to islands of lymphocytes separated by expanded medullary and subcapsular sinuses and bare scaffolds of the lymph node cortex. In these remaining islands of lymphoid tissue the numbers of cells positive for SIV RNA by in situ hybridization is large as shown in this figure and in figure 6 at a higher magnification.

observed (fig. 6) which appeared to correlate with the number of tingible body macrophages. The cells which contained SIV RNA in these nodes were both small lymphocytic cells and larger cells compatible with macrophages. Cells compatible with medium and small lymphocytes comprised less than 50% of the in situ hybridization-positive cells in these cases. No signal was noted in cases with prominent angiogenesis. In some monkeys a large number of multinucleate cells was observed in the medullary and subcapsular sinuses of the lymph nodes (fig. 7). Multinucleate cells were not seen as frequently in the parenchyma as they were in the sinuses (fig. 8). Macrophages in the subcapsular sinuses in cases without giant cells were rarely positive.

In situ hybridization is a useful technique for the study of the pathogenesis of SIV-induced immunodeficiency in rhesus monkeys. The results presented in this paper indicate that SIV RNA is abundant in the brains of SIV-infected monkeys and are consistent with previous electron microscopic observations of SIV within phagosomes of brain macrophages [21, 30]. The

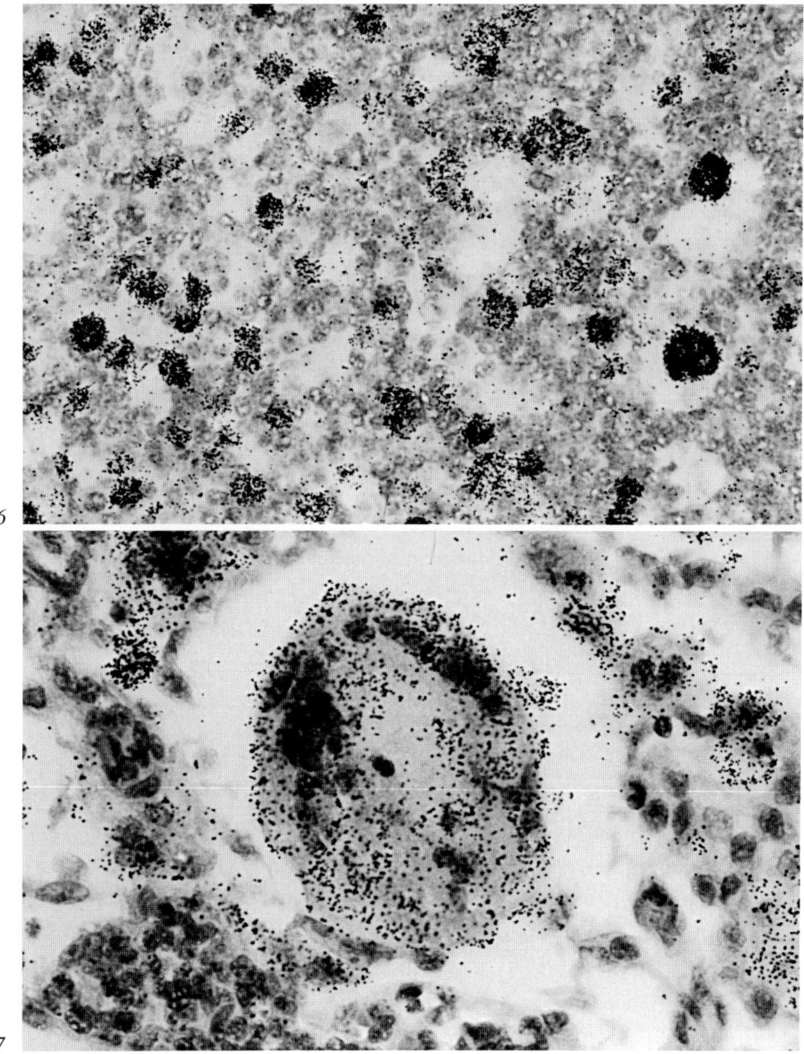

Fig. 6. The number of cells positive for SIV RNA by in situ hybridization are extensive. Smaller cells compatible with lymphocytes have fewer number of grains per cell when compared to the larger cells which are compatible with macrophages. In this figure the amount of SIV RNA varies widely from cell to cell.

Fig. 7. Large multinucleate cells are observed in the lymph nodes of some rhesus monkeys inoculated with SIV. This high magnification of such a cell within a medullary sinus illustrates the presence of SIV RNA detected by in situ hybridization.

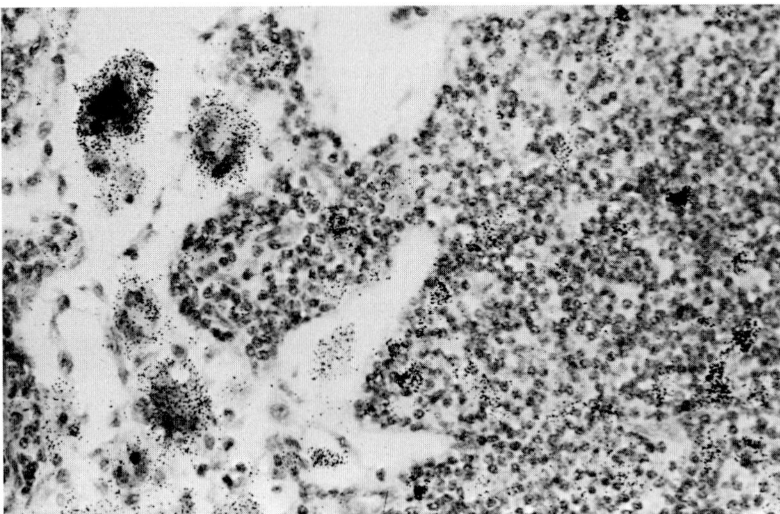

Fig. 8. Multinucleate cells are present in the sinus adjacent to a parenchymal area in which only SIV-positive mononuclear cells are visible. In cases in which giant cells were not observed, the numbers of positive cells in the sinuses were very low.

lesions present in the rhesus monkey are morphologically similar to lesions observed in HIV-infected human brains with the exception that myelin pallor and astrocytosis are less frequently observed [32]. The SIV model system offers an ability to investigate the virus or host factors which result in brain lesions and to intervene early in the disease to study their genesis.

In lymphoid tissues in situ hybridization with tritium-labelled probes clearly identifies cells which contain SIV RNA at various stages of lymphoid hyperplasia and lymphoid depletion. The number and distribution of cells containing SIV RNA appears to correlate in a predictable manner with the progressive morphological stages leading to lymphoid depletion. Numbers of infected cells are rare during follicular hyperplasia and numerous during stages of follicular depletion with paracortical atrophy and tingible body macrophage activity. These findings agree with the histologic changes and the in situ hybridization results seen in lymph nodes from HIV-infected humans with the exception that a large number of cells containing HIV RNA have not been reported in human lymph nodes [4, 34]. The reason for this difference may be due, in part, to the fact that human AIDS patients usually die with lymph nodes in profound lymphocytic depletion [4, 11] and that the experimentally inoculated macaques in this study die earlier in the course of their disease since they do not receive any therapy. The difference in the number of cells containing SIV RNA observed during stages of lymphadenopathy compared with later stages of lymphoid involution does suggest that widespread viral replication in germinal centers may not be the cause of the progressive generalized lymphadenopathy. In both humans and monkeys the presence of viral structural proteins in the absence of significant virus or viral RNA [4, 31, 34–38] in germinal centers suggests that antigen-antibody complexes associated with follicular dendritic cells and not direct viral replication may stimulate follicular hyperplasia. The reasons for the dramatic infection of macrophages and lymphocytes during later stages of lymphoid involution are not clear but certainly indicate that SIV has in the terminal stage of disease greatly accelerated its ability to evade the host defenses.

The SIV-infected rhesus monkey will undoubtedly play a major role in the study of AIDS pathogenesis and in the development of new anti-AIDS drugs and vaccine strategies. The molecular biology of SIV and HIV are so similar that antiviral agents developed against SIV will likely be directly applicable to HIV. The close immunologic and physiologic relationship of nonhuman primates to humans makes the rhesus monkey system an important model to study virus host interactions which lead to immunodeficiency. We are continuing our in situ hybridization studies to include the gastrointestinal and reproductive systems and hope that these studies will increase our knowledge of SIV cellular and tissue tropisms.

References

1 Alter, H.; Eichberg, J.; Masur, H.; Saxinger, W.; Gallo, R.: Transmission of HTLV-III infection from human plasma to chimpanzees: An animal model for AIDS. Science *226:* 549–552 (1984).

2 Baskin, G.; Martin, L.; Rangan, S.; Gormus, B.; Murphey-Corb, M.; Wolf, R.; Soike, K.: Transmissible lymphoma and simian acquired immunodeficiency syndrome in rhesus monkeys. JNCI *77:* 127–139 (1986).

3 Benveniste, R.; Arthur, L.; Tsai, C.; Sowder, R.; Copeland, T.; Henderson, L.; Oroszlana, S.: Isolation of a lentivirus from a macaque with lymphoma: Comparison with HTLV-III/LAV and other lentiviruses. J. Virol. *60:* 483–490f (1986).

4 Biberfeld, P.; Chayt, K.; Marselle, L.; Biberfeld, G.; Gallo, R.; Harper, M.: HTLV-III expression in infected lymph nodes and relevance to pathogenesis of lymphadenopathy. Am. J. Pathol *125:* 436–442 (1986).

5 Brigati, D.; Myerson, D.; Leary, J.; Spalholz, B.; Travis, S.; Fong, C.; Hsiung, G.; Ward, D.: Detection of viral genomes in cultured cells and paraffin-embedded tissue sections using biotin-labeled hybridization probes. Virology *126:* 32–50 (1983).
6 Chakrabarti, L.; Guyader, M.; Alizon, M.; Daniel, M.; Desrosiers, R.; Tiollais, P.; Sonigo, P.: Sequence of simian immunodeficiency virus from macaque and its relationship to other human and simian retroviruses. Nature *328:* 543–547 (1987).
7 Chalifoux, L.; Ringler, D.; King, N.; Sehgal, P.; Desrosiers, R.; Daniel, M.; Letvin, N.: Lymphadenopathy in macaques experimentally infected with the simian immunodeficiency virus (SIV). Am. J. Pathol. *128:* 104–110 (1987).
8 Chalifoux, L.; King, N.; Letvin, N.: Morphologic changes in lymph nodes of macaques with an immunodeficiency syndrome. Lab. Invest. *51:* 22–26 (1984).
9 Daniel, M.; Letvin, N.; King, N.; Kannagi, M.; Sehgal, P.; Hunt, R.; Kanki, P.; Essex, M.; Desrosiers, R.: Isolation of T-cell tropic HTLV-III-like retrovirus from macaques. Science *228:* 1201–1204 (1985).
10 Daniel, M.; Letvin, N.; Sehgal, P.; Hunsmann, G.; Schmidt, D.; King, N.; Desrosiers, R.: Long-term persistent infection of macaque monkeys with the simian immunodeficiency virus. J. Gen. Virol. *68:* 3183–3189 (1987).
11 Diebold, J.; Marche, C.; Audouin, J.; Aubert, J.; Le Tourneau, A.; Bouton, C.; Reynes, M.; Wizniak, J.; Capron, F.; Tricottet, V.: Lymph node modification in patients with the acquired immunodeficiency syndrome (AIDS) or with AIDS-related complex (ARC). Pathol. Res. Pract. *180:* 590–611 (1985).
12 Filice, G.; Cereda, P.; Varnier, O.: Infection of rabbits with human immunodeficiency virus. Nature *335:* 366–369 (1988).
13 Fultz, P.; McClure, H.; Anderson, D.; Swenson, R.; Anand, R.; Srinivasan, A.: Isolation of a T-lymphocyte retrovirus from naturally infected sooty mangabey monkeys *(Cerocebus atys)*. Proc. Natl. Acad. Sci. USA *83:* 5286–5290 (1986).
14 Fultz, P.; McClure, H.; Swenson, R.; McGrath, C.; Brodie, A., et al.: Persistent infection of chimpanzees with human T-lymphotropic virus type III/lymphadenopathy-associated virus: a potential model for acquired immunodeficiency syndrome. J. Virol. *58:* 116–124 (1986).
15 Gajdusek, D.; Amyx, H.; Gibbs, C., Jr.; Asher, D.; Rodgers-Johnson, P., et al.: Infection of chimpanzees by human T-lymphotropic retroviruses in brain and other tissues from AIDS patients (letter). Lancet *i:* 55–56 (1985).
16 Gajdusek, D.; Amyx, H.; Gibbs, C., Jr.; Asher, D.; Yanagihara, R., et al.: Transmission experiments with human T-lymphotropic retroviruses and human AIDS tissue (letter). Lancet *i:* 1415–1416 (1984).
17 Kanki, P.; McLane, M.; King, N.; Letvin, N.; Hunt, R.; Sehgal, P.; Daniel, M.; Desrosiers, R.; Essex, M.: Serologic identification and characterization of a macaque T-lymphotropic retrovirus closely related to HTLV-III. Science *228:* 1199–1201 (1985).
18 Kannagi, M.; Yetz, J.; Letvin, N.: In vitro growth characteristics of simian T-lymphotropic virus type III. Proc. Natl. Acad. Sci. USA *82:* 7053–7057 (1985).
19 Kannagi, M.; Kiyotaki, M.; Desrosiers, R.; Reimann, K.; King, N.; Waldron, L.; Letvin, N.: Humoral immune responses to T cell tropic retrovirus simian T lymphotropic virus type III in monkeys with experimentally induced acquired immune deficiency-like syndrome. J. Clin. Invest. *78:* 1229–1236 (1986).
20 King, N.: Simian models of acquired immunodeficiency syndrome (AIDS). A review. Vet. Pathol. *23:* 345–353 (1986).

21 Letvin, N.; Daniel, M.; Sehgal, P.; Desrosiers, R.; Hunt, R.; Waldron, L.; MacKey, J.; Schmidt, D.; Chalifoux, L.; King, N.: Induction of AIDS-like disease in macaque monkeys with T-cell tropic retrovirus STLV-III. Science *230:* 71–73 (1985).
22 Lowenstine, L.; Pedersen, N.; Higgins, J.; Pallis, K.; Uyeda, A.; Marx, P.; Lerche, N.; Munn, R.; Gardner, M.: Seroepidemiologic survey of captive Old-World primates for antibodies to human and simian retroviruses, and isolation of a lentivirus from sooty mangabeys *(Cerocebus atys)*. Int. J. Cancer *38:* 563–574 (1986).
23 Maniatis, T.; Fritsch, E.; Sambrook, J.: Molecular cloning: A laboratory manual; in Molecular Cloning – A Laboratory Manual (Cold Spring Harbor Laboratory, Cold Spring Harbor, N.Y. 1982).
24 Murphey-Corb, M.; Martin, L.; Rangan, S.; Baskin, G.; Gormus, B.; Wolf, R.; Andes, W.; West, M.; Montelaro, R.: Isolation of an HTLV-III-related retrovirus from macaques with simian AIDS and its possible origin in asymptomatic mangabeys. Nature *321:* 435–437 (1986).
25 Namikawa, R.; Kaneshima, H.; Lieberman, M.; Weissman, I.; McCune, J.: Infection of the SCID-Hu mouse by HIV-1. Science *242:* 1684–1686 (1988).
26 Niedrig, M.; Rabanus, J.; Stehr, J.; Gelderblom, H.; Pauli, G.: Monoclonal antibodies directed against human immunodeficiency virus (HIV) gag proteins with specificity for conserved epitopes in HIV-1, HIV-2 and simian immunodeficiency virus. J. Gen. Virol. *69:* 2109–2114 (1988).
27 Rigby, P.; Dreckmann, M.; Rhodes, C.; Berg, P.: Labeling deoxyribonucleic acid to high specific activity in vitro by nick translation with DNA polymerase I. J. Mol. Biol. *113:* 237–251 (1977).
28 Ringler, D.; Murphy, G.; King, N.: An erythematous maculopapular eruption in macaques infected with an HTLV-III-like virus (STLV-III). J. Invest. Dermatol. *87:* 674–677 (1986).
29 Ringler, D.; Hancock, W.; King, N.; Letvin, N.; Daniel, M.; Desrosiers, R.; Murphy, G.: Immunophenotypic characterization of the cutaneous exanthem of SIV-infected rhesus monkeys. Am. J. Pathol. *126:* 199–207 (1987).
30 Ringler, D.; Hunt, R.; Desrosiers, R.; Daniel, M.; Chalifoux, L.; King, N.: Simian immunodeficiency virus-induced meningoencephalitis: Natural history and retrospective study. Ann. Neurol. *23:* suppl., pp. S101–S107 (1988).
31 Ringler, D., Wyand, M.; Walsh, D.; Mackey, J.; Chalifoux, L.; Popovic, M.; Minassian, A.; Sehgal, P.; Daniel, M.; Desrosiers, R.; King, N.: Cellular localization of simian immunodeficiency virus (SIV) in lymphoid tissues. I. Immunohistochemistry and electromicroscopy. Am. J. Pathol. *134:* 373–383 (1989).
32 Sharer, L.; Baskin, G.; Cho, E.; Murphy-Corb, M.; Blumberg, B.; Epstein, L.: Comparison of simian immunodeficiency virus and human immunodeficiency virus encephalidities in the immature host. Ann. Neurol. *23:* suppl., pp. S108–S112 (1988).
33 Singer, R.; Lawrence, J.; Villnave, C.: Optimization of in situ hybridization using isotopic and non-isotopic detection methods. BioTechniques *4:* 230–250 (1986).
34 Sun, N.; Shapshak, P.; Schmid, P.; Hsu, M.; Beall, G.; Imagawa, D.: Comparison of HIV antigens and nucleic acids in biopsied lymph nodes from patients at risk for HIV infection (Abstract). Lab. Invest. *58:* 89 (1988).
35 Tenner-Racz, K.; Racz, P.; Bofill, M.; Schulz-Meyer, A.; Dietrich, M.; Kern, P.; Weber, J.; Pinching, A.; Veronese-Dimarzo, F.; Popovic, M.; Klatzmann, D.; Gluckman, J.; Janossy, G.: HTLV-III/LAV viral antigens in lymph nodes of homosexual

men with persistent generalized lymphadenopathy and AIDS. Am. J. Pathol. *123:* 9–15 (1986).

36 Tenner-Racz, K.; Racz, P.; Kern, P.; Dietrich, M.; Ramsauer, J.; Popovic, M.; St. Georg, A.: Germinal centers in the persistence of HIV infection (Abstract). Lab. Invest. *58:* 93 (1988).

37 Wyand, M.; Ringler, D.; Naidu, Y.; Mattmuller, M.; Daniel, M.; Desrosiers, R.; King, N.: Detection of simian immunodeficiency virus in macaque lymph nodes with a SIV-mac envelope probe. J. Med. Primatol. *18:* 209–215 (1989).

38 Wyand, M.; Ringler, D.; Naidu, Y.; Mattmuller, M.; Chalifoux, L.; Sehgal, P.; Daniel, R.; King, N.: Cellular localization of simian immunodeficiency virus in lymphoid tissues. II. In situ hybridization. Am. J. Pathol. *134:* 385–393 (1989).

Michael S. Wyand, DVM, PhD, EG&G Mason Research Institute, 57 Union Street, Worcester, MA 01608 (USA)

Racz P, Haase AT, Gluckman JC (eds): Modern Pathology of AIDS and Other Retroviral Infections. Basel, Karger, 1990, pp 155–160

Epstein-Barr Virus DNA in Oral Mucosa of HIV-Infected Patients[1]

R.-P. Henke[a], *M. Winzer*[b], *T. Löning*[a]

[a] Institute of Pathology, University of Hamburg; [b] Department of Dermatology, University of Lübeck, FRG

The oral mucosa is frequently involved in the broad spectrum of symptoms caused by infection with the human immunodeficiency virus (HIV). In 1984, Greenspan et al. [1] described a hitherto unknown lesion of the tongue mucosa, which was associated with the acquired immunodeficiency syndrome (AIDS) and which they denoted as 'hairy leukoplakia' (HL) according to the microscopic appearance of the superficial parakeratotic cells. Initially, these lesions were thought to be a peculiar manifestation of candidiasis. After it was shown that they could be therapeutically influenced by virustatic medication, a virus of the herpes group was hypothetically proposed as the etiologic agent. Finally, this supposition was advocated by the detection of Epstein-Barr virus (EBV) nucleic acids and structural proteins in lesions of HL [2].

In the present paper, we describe the histologic and cytologic appearance of EBV-associated lesions of the oral mucosa including HL and correlate it with the presence of EBV-infected epithelial cells ascertained by in situ hybridization.

Materials and Methods

Patients

Punch biopsies of 34 patients with verified (ELISA, Western blot) HIV infection were investigated by in situ hybridization for the presence of EBV nucleic acid sequences. HL had been diagnosed in 17 cases. In addition to in situ hybridization, frozen or paraffin-embedded tissue specimens were routinely processed for hematoxylin/eosin and PAS staining.

[1]Supported by the Deutsche Forschungsgemeinschaft (Lo285/2–3) and the Hamburger Stiftung zur Förderung der Krebsbekämpfung.

In situ Hybridization

In situ hybridization was performed as detailed elsewhere [3–5]. Briefly, paraffin sections were adhered to aminoalkylsilane-treated glass slides [6], dewaxed in xylene, incubated with pronase (1 mg/ml), dehydrated through a graded series of alcohol and air dried. For hybridization, each section was covered with 20 μl of the following hybridization solution: 2 × SSC (1 × SSC ≡ 0.15 *M* NaCl, 0.015 *M* trisodium citrate; pH 7.2), 20% (v/v) deionized formamide, 10% (w/v) dextran sulfate, 0.1 mg/ml herring sperm DNA, and 1.0 μ/ml biotinylated EBV DNA (Enzo, New York; prepared from the 3 kb BamHI 'V' fragment cloned into the BamHI site of pBR322). After denaturation by heating to 90 °C (for 10 min), hybridization was allowed to take place at 37 °C overnight. Consecutively, sections were washed twice for 10 min in 1 × SSC, 45% formamide at 37 °C, followed by three washes for 5 min each in double strength SSC at room temperature. For detection of hybridized probes, sections were subsequently covered with (a) rabbit antibiotin antibodies (Enzo); (b) biotinylated goat anti-rabbit antiserum (Dianova, Hamburg, FRG), and (c) a streptavidin-alkaline phosphatase (BRL) or streptavidin-colloidal gold (5 nm) (Janssen, Beerse, Belgium) conjugate visualized for light microscopy by incubation with nitro blue tetrazolium/bromochloroindolyl phosphate or a silver enhancement procedure (IntenSE, Janssen).

Controls included replacement of the EBV probe by biotinylated cytomegalovirus (CMV; Enzo) or human papillomavirus (HPV) probes (types 6, 11, 13, 16 or 18 harboring plasmids were generously provided by Drs. H. zur Hausen and L. Gissmann, Deutsches Krebsforschungszentrum, Heidelberg).

Results and Discussion

EBV DNA could be detected in 17 oral mucosal specimens of 34 HIV-infected patients. Thirteen of these 17 biopsies were classified as HL according to clinical and light microscopical criteria. All EBV-positive tissue specimens were characterized by particular changes. The basal cell layer was not significantly enlarged and usually confined to a single row of keratinocytes. Above the basal cell layer an increasing degree of intracellular edema (ballooning) of spinous cells was noted (fig. 1, 2). This feature was most striking at the top of the stratum spinosum just beneath a small zone of three to four layers of flattened parakeratinized epithelial cells. Due to the spongy aspect of the most superficial spinous cells, the parakeratotic surface was often abruptly delineated from the stratum spinosum. Cytologically, a nuclear chromatolysis and so-called ground-glass nuclei were noted. In addition, basophilic nuclear inclusions could be observed.

EBV DNA was demonstrated preferentially in nuclei of spongiotic keratinocytes within the upper stratum spinosum (fig. 3). Neither basal epithelial cells nor stromal cells showed positive hybridization signals. In contrast to previous reports [1, 2], in our series of cases EBV DNA production was not confined to tongue epithelium (2 positive cases of gingival mucosa) and, in

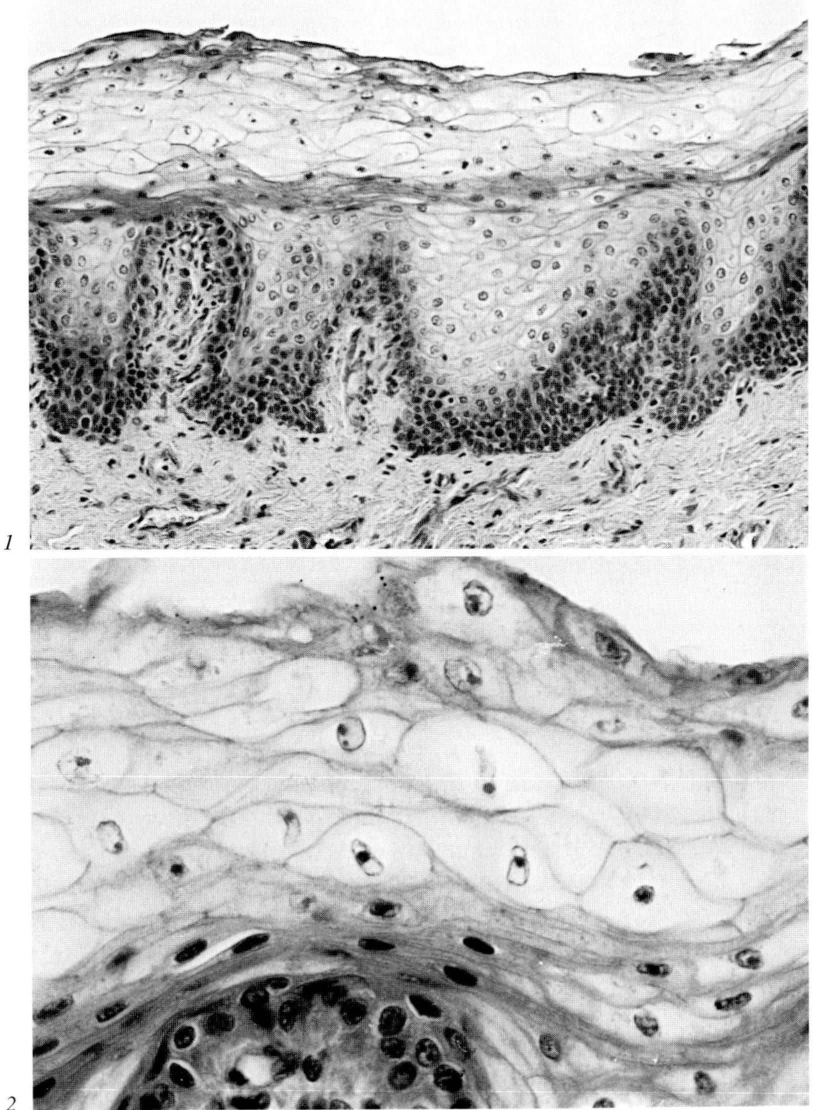

Fig. 1. HL: flat epithelium with pronounced broadening of the upper stratum spinosum. HE. ×90.

Fig. 2. Same case. Marked edema and ballooning of spinous cells associated with degeneration of nuclei. HE. ×435.

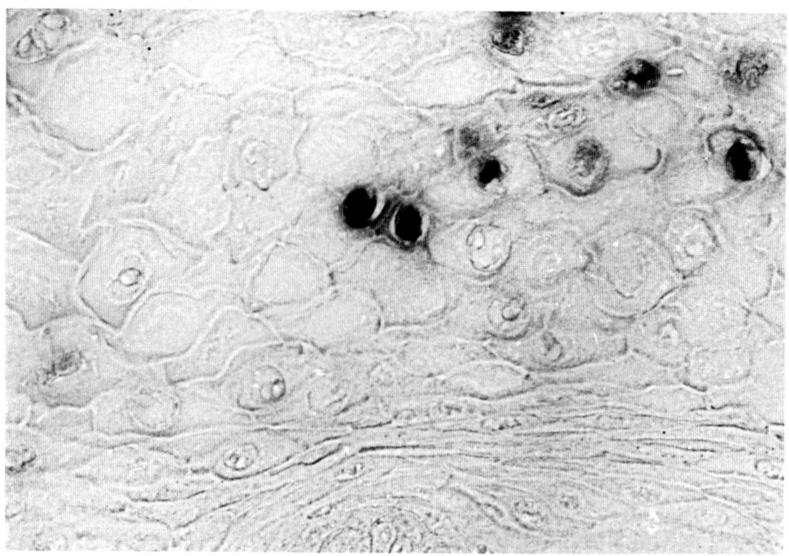

Fig. 3. Same case. In situ hybridization with EBV DNA probe revealing intensive labelling of nuclei within fields of edematous spinous cells. Alkaline phosphatase technique. ×435.

addition, was not restricted to clinically apparent cases of HL, as was evidenced by 4 EBV-positive cases of normal appearing mucosa. In 2 cases morphological changes consistent with EBV infection as well as EBV DNA were detected in HIV-infected, but otherwise asymptomatic patients, according to CDC stage II [7]. Positive results with the CMV probe were not obtained. In addition, under stringent washing conditions all tissue specimens were negative for the applied HPV probes of types 6, 11, 13, 16 and 18.

AIDS patients are frequently coinfected with EBV, with up to 96% showing serologic evidence of active infection [8]. Experimentally, mutual effects of both viruses could be observed. Birx et al. [9] showed that T cells of patients with AIDS or AIDS-related disorders are defective in suppressing autologous B cells infected with EBV, offering a possible explanation for the abnormally high numbers of EBV-infected B cells found in the circulation of these patients. On the other hand, an EBV immediate-early gene product, BamHI *MLFl,* stimulates expression of genes linked to the HIV promoter by a transacting mechanism [10].

As was shown by Young et al. [11] and Sixbey et al. [12], the serum complement receptor, C3d, which serves as a functional receptor for EBV, is expressed on epithelial cells of the oropharynx. Thereby, keratinocytes of the oral mucosa are susceptible to infection with EBV by the same route as the classic target cell of EBV, the B lymphocyte. The hypothesis that EBV is involved in the etiology of HL was strongly advocated by the detection of EBV nucleic acids and structural proteins in lesions of HL by Greenspan et al. [2] in 1985. Later, these authors stressed the prognostic relevance of HL in the course of the development of full-blown AIDS, which was observed within 31 months after clinical manifestation of HL in 83% of their patients [13].

The results of our study substantiate the assumption that HL represents a characteristic tissue reaction with distinct cytopathic effects in response to an EBV infection, which is unusually productive in association with the underlying HIV infection and which can be diagnosed by in situ hybridization in a high percentage of cases.

References

1 Greenspan, D.; Greenspan, J.S.; Conant, M.; Petersen, V.; Silverman, S.; de Souza, Y.: Oral 'hairy leucoplakia' in male homosexuals: evidence of association with both papillomavirus and a herpes-group virus. Lancet *ii:* 831–834 (1984).

2 Greenspan, J.S.; Greenspan, D.; Lennette, E.D.; Abrams, D.I.; Conant, M.A.; Petersen, V.; Freese, U.K.: Replication of Epstein-Barr virus within the epithelial cells of oral 'hairy' leukoplakia, an AIDS-associated lesion. N. Engl. J. Med. *313:* 1564–1571 (1985).

3 Milde, K.; Löning, T.: Detection of papillomavirus DNA in oral papillomas and carcinomas: application of in situ hybridization with biotinylated HPV 16 probes. J. Oral Pathol. *15:* 292–296 (1986).

4 Henke, R.-P.; Milde-Langosch, K.; Loening, T.; Strømme Koppang, H.: Human papillomavirus type 13 and focal epithlial hyperplasia of the oral mucosa: DNA hybridization on paraffin-embedded specimens. Virchows Arch. (A) *411:* 193–198 (1987).

5 Löning, T.; Henke, R.-P.; Reichart, P.; Becker, J.: In situ hybridization to detect Epstein-Barr virus DNA in oral tissues of HIV-infected patients. Virchows Arch. (A) *412:* 127–133 (1987).

6 Rentrop, M.; Knapp, B.; Winter, H.; Schweizer, J.: Aminoalkyl-silane-treated glass slides as support for in situ hybridization of keratin cDNAs to frozen tissue sections under varying fixation and pretreatment conditions. Histochem. J. *18:* 271–276 (1986).

7 Centers for Disease Control: Classification system for human T-lymphotropic virus type III/lymphadenopathy-associated virus infections. MMWR *35:* 334–339 (1986).

8 Fauci, A.S.; Macher, D.L.; Longo, D.L.; Lane, H.C.; Rook, A.H.; Masur, M.; Gelman, E.P.: Acquired immunodeficiency syndrome: Epidemiologic, clinical, immunologic, and therapeutic considerations. Ann. Intern. Med. *100:* 92–106 (1984).

9 Birx, D. L.; Redfield, R. R.; Tosata, G.: Defective regulation of Epstein-Barr virus infection in patients with acquired immunodeficiency syndrome (AIDS) or AIDS-related disorders. N. Engl. J. Med. *314:* 874–879 (1986).

10 Kenney, S.; Kamine, J.; Markovitz, D.; Fenrick, R.; Pagano, J.: An Epstein-Barr virus immediate-early gene product transactivates gene expression from the human immunodeficiency virus long terminal repeat. Proc. Natl. Acad. Sci. USA *85:* 1652–1656 (1988).

11 Young, L. S.; Sixbey, J. W.; Clark, D.; Rickinson, A. B.: Epstein-Barr virus receptors on human pharyngeal epithelia. Lancet *i:* 240–242 (1986).

12 Sixbey, J. W.; Nedrud, J. G.; Raab-Traub, N.; Hanes, R. A.; Pagano, J. S.: Epstein-Barr virus replication in oropharyngeal epithelial cells. N. Engl. J. Med. *310:* 1225–1230 (1984).

13 Greenspan, D.; Greenspan, J. S.; Hearst, N. G.; Pan, L.-Z.; Conant, M. A.; Abrams, D. I.; Hollander, H.; Levy, J. A.: Relation or oral hairy leukoplakia to infection with the human immunodeficiency virus and the risk of developing AIDS. J. Infect. Dis. *155:* 475–480 (1987).

Dr. R.-P. Henke, Institute of Pathology, University of Hamburg, Martinistrasse 52, D–2000 Hamburg 20 (FRG)

Racz P, Haase AT, Gluckman JC (eds): Modern Pathology of AIDS and
Other Retroviral Infections. Basel, Karger, 1990, pp 161–176

Ultrastructural Aspects of Lymphadenopathy Induced by the Feline Leukemia Virus[1]

Evidence for Intrafollicular Virus Replication and Spread of Infection through the Lymphatics

Klara Tenner-Racz[a], *Paul Racz*[b], *Luis M. Taveres*[c],
Heidemarie Schmidt[b], *Fernando de Noronha*[c,2]

[a]Department of Haematology, Allgmeines Krankenhaus St. Georg, Hamburg,
[b]Department of Pathology and Körber Laboratory for AIDS Research,
Bernhard Nocht Institute for Tropical Medicine, Hamburg, FRG;
[c]New York State College of Veterinary Medicine, Cornell University,
Ithaca, N.Y., USA

Feline leukemia virus (FeLV), discovered by Jarrett et al. [19, 20], is widespread among domestic cats. It was named for its original association with leukemogenesis. However, it was soon recognized that not only lymphoproliferative diseases develop in infected animals. Aplastic anemia or severe progressive immune hyporesponsiveness accompanied by thymic atrophy and lymphoid depletion of lymph nodes has been documented in many infected cats [1, 12–18, 33]. The immunosuppression makes these animals extremely susceptible to opportunistic infections which are the major cause of death.

The acquired immunodeficiency syndrome (AIDS) has greatly stimulated interest in animal diseases caused by retroviruses. Although none of the existing animal models can completely mimic the disease caused by the human immunodeficiency virus (HIV), they can facilitate the analysis of the patho-

[1]This work was sponsored by BMFT (Bundesministerium für Forschung und Technologie), Bonn; Körber-Stiftung, Hamburg, and by contract S850-25, NCI/Frederick Cancer Research Facility/Programm Resources Inc., National Cancer Institute Grant 1RO1-CA37742-01A1.

[2]The authors thank Ms. Brigitte Kröger, Angela Pries and Ulrike Fritzsche for excellent technical assistance.

genesis and be helpful in testing experimental vaccines or new therapy regimens, particularly those with antiviral effects.

The FeLV-induced immunodeficiency disease is an attractive model because it shares many similarities with HIV infection. Cats infected by FeLV may develop severe diarrhea, weight loss, neutropenia, lymphopenia, generalized lymphadenopathy, cutaneous anergy, reduced lymphocyte response to mitogens, impaired humoral response to neoantigens, e.g., synthetic polypeptides. Histologic examinations of lymph nodes have shown a follicular hyperplasia in the early phase of infection. During the course of the disease, germinal centers (GC) disintegrate and the process ends with lymphocyte-depleted lymph nodes [for review, see 14, 16]. Thus, the lymph node histopathology appears essentially the same as in HIV infection.

Immunohistochemical analysis has demonstrated group-specific FeLV antigen in GC of lymph nodes from infected cats. The findings indicate that viral replication occurs at this location [16, 18, 39]. Similar immunohistochemical results can be obtained when monoclonal antibodies raised against *gag* proteins of HIV are applied on frozen sections of lymph nodes from patients with persistent generalized lymphadenopathy (PGL) induced by the causative agent of AIDS [4, 30, 32, 34, 46–48, 51, 52]. Through electron microscopy the presence of HIV between the cell processes of follicular dendritic cells (FDC) [2, 3, 6, 34, 45, 46, 49–52, 55] and virus replication involving different cell types such as FDC [2, 3, 52], lymphocytes and macrophages [52] has been demonstrated in PGL. Using in situ hybridization, cells with mRNA of HIV can be found mainly in GC [5, 35, 50]. Ultrastructural analyses on repeated biopsy specimens have shown that HIV can persist in GC for long periods of time [51, 52].

Electron microscopic investigations of GC in FeLV infection have not been performed in the past. The amount and localization of virions and the identification of virus-producing cells are very important and contribute to our understanding of the pathogenesis of the disease. The kinetics of FeLV infection, under experimental conditions, are well characterized [1, 7, 9, 12–19, 26, 27, 37–40]. This feline system has been used to evaluate the efficiency of different antiretroviral therapies [17, 43, 44]. To monitor the effectiveness of such therapeutic approachment or chemoprophylaxis, ultrastructural analysis of GC would yield important information. The aim of the present study was directed toward the intrafollicular localization of FeLV and the identification of cells involved in virus production. Special attention was devoted to the question of similarities to and differences from alterations as seen in HIV-induced follicular hyperplasia.

Materials and Methods

Experimental Animals. Cats used were obtained from the Cornell Specific Pathogen Free (SPF) Cat Colony [26] and housed in Horsfall cages equipped with double air filtration.

Virus Strain. The FeLV strain used in this study was biologically cloned from the Rickard strain as reported [43]. The immunosuppressive characteristics of this virus strain have been defined [manuscript in preparation].

Experimental Infection. Five 6-to-8-week-old SPF kittens were infected by subcutaneous inoculation of 2×10^3 focus-inducing units (FIU) of FeLV. One animal was maintained as an age-matched uninfected control. Blood samples were drawn every 14 days to evaluate viremia and serum antibody status. All animals were euthanized 6 months after virus inoculation. Mesenteric, cervical, retropharyngeal, mediastinal and axillary lymph nodes were subsequently collected for electron microscopical analysis.

Virus Titration. Viremia levels and other virus titrations were carried out with the 81C cell assay previously described [27].

Assay for Evaluation of Anti-FeLV Antibodies. Neutralization using heat-inactivated serum samples (undiluted, 1:10, 1:20 and 1:40) from all animals was performed biweekly as described [43]. Serum dilutions causing a focus reduction of greater than 50%, as compared with SPF serum control plates, were considered positive.

Electron Microscopy. Portions of each lymph node were fixed in 4% glutaraldehyde (cacodylate buffer, pH 7.2), postfixed in 1% OsO_4 (cacodylate buffer, pH 7.2), dehydrated in graded acetone and routinely embedded in Araldite. Semithin sections were cut using glass knives, then stained with methylene blue-azure II stain and examined with a light microscope for the presence of GC. Upper areas of GC and parts of marginal sinus were selected for electron microscopical investigations. Ultrathin sections were cut using a diamond knife, and subsequently stained with uranyl acetate and lead citrate before examining with a Philips 201 electron microscope operated at 80 kV.

Results and Discussion

Viremia and neutralizing antibody levels <1:5 developed in all cats 2–4 weeks after virus inoculation. The viremia titers ranged from 3×10^3 to 5×10^4 FIU/ml of blood and persisted with a uniform titer of $1\text{–}3 \times 10^3$ FIU/ml throughout the experimental period despite detectable virus-neutralizing antibodies.

Electron microscopical examination revealed a high amount of cell-free C-type virus particles in GC of all lymph nodes obtained from infected cats. Characteristically, the virions were located between the cell processes of FDC. The dendrites were slightly swollen and their number was elevated (fig. 1, 2).

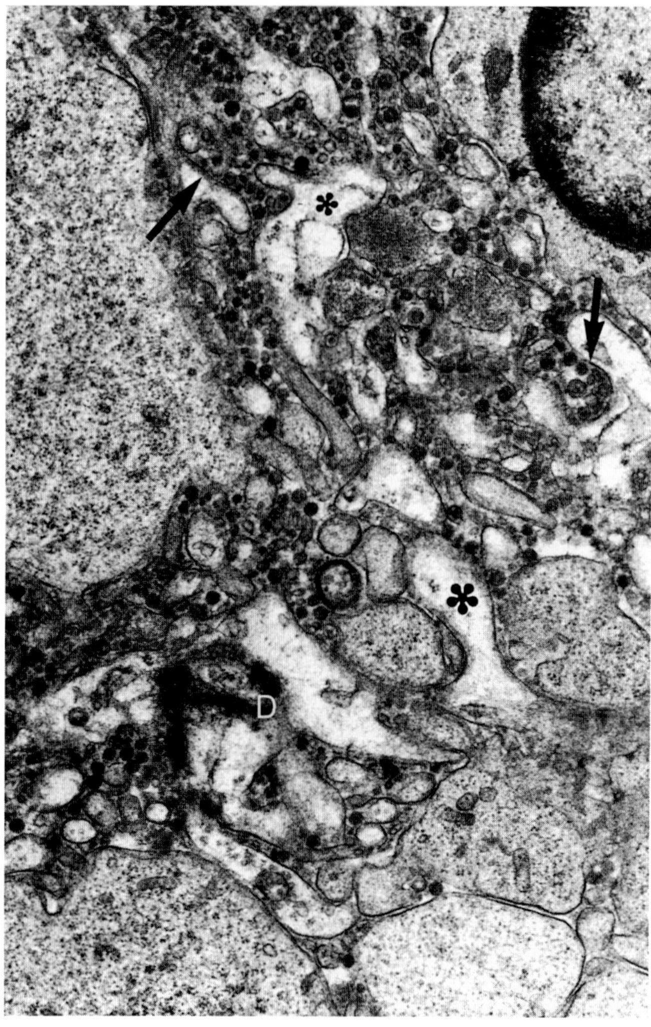

Fig. 1. Increased number of dendrites in FeLV-induced follicular hyperplasia. Many dendrites are swollen (asterisks). There are a lot of FeLV in the interdendritic extracellular spaces (arrows). D = desmosome. × 11,700.

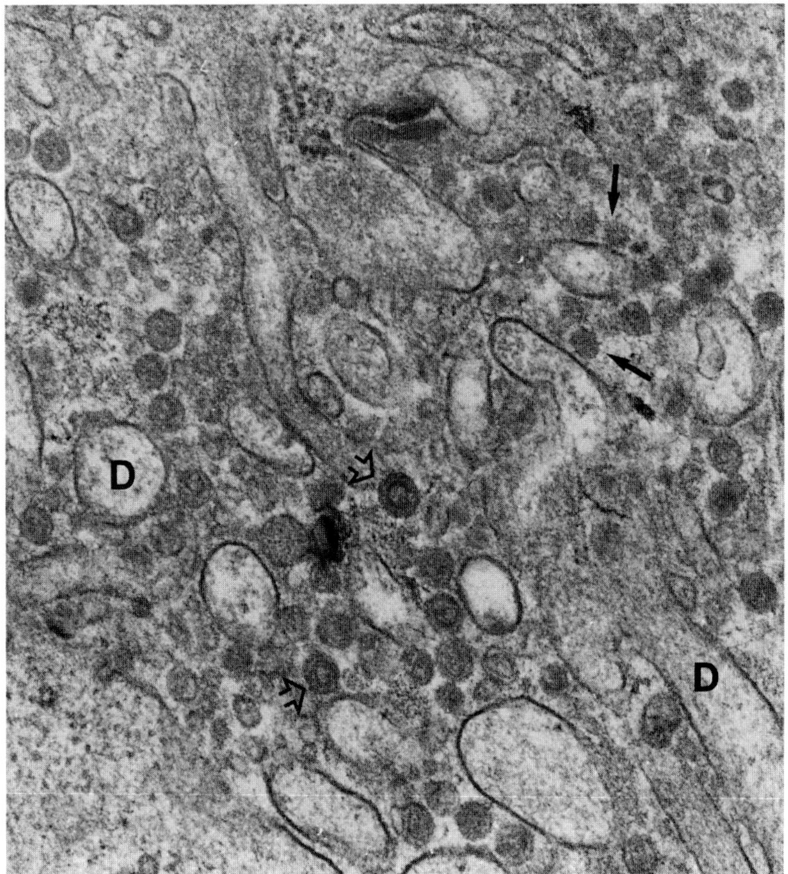

Fig. 2. Complete (open arrows) and incomplete (arrows) virus particles between the cytoplasmic processes of FDC (D). ×23,400.

None or only a few virions were seen between lymphoid cells where cell processes of FDC were not present (fig. 3).

The diameter of virus particles avaraged 100 nm and many of them contained a well-defined core. In addition, there were numerous moderately electron-dense round structures without recognizable internal core and surrounded by a membrane similar to that of FeLV (fig. 2). Such structures were not found in the control animal or in regions of infected lymph nodes where complete virus particles were absent. We assume that these incomplete parti-

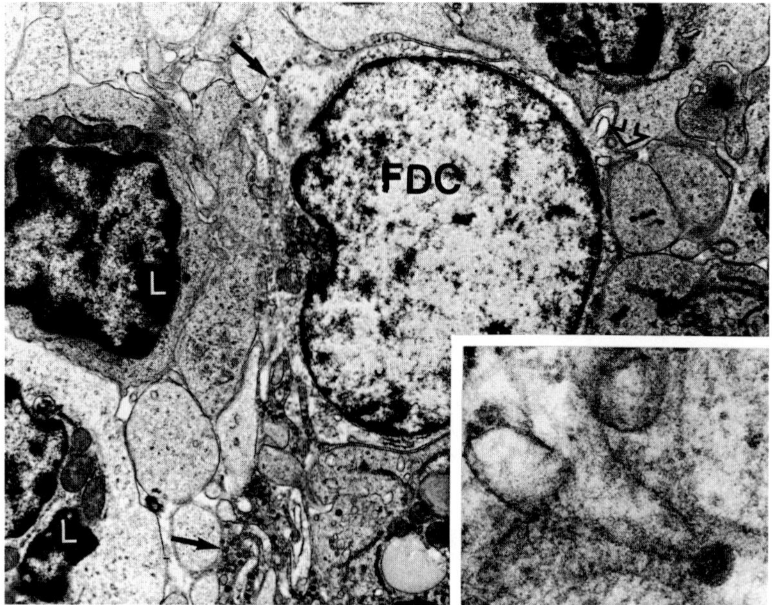

Fig. 3. Accumulation of FeLV around a FDC (arrows). No virions are seen between lymphoid cells (L) where dendrites are absent. Open arrow and inset show a budding FeLV from the cell surface of a FDC. × 5,450. Inset: × 23,400.

cles represent defective FeLV and that they are generated during viral replication.

The distribution of FeLV seen in the present study is identical with that of HIV in PGL where typical virions with lentiviurs morphology [2, 3, 6, 30, 34, 45, 46, 48, 51, 52, 55] and incomplete particles [49, 52] have been described. Using double immunolabeling with monoclonal antibodies against FDC and *gag* proteins of HIV, the staining shows the close association of viral proteins with FDC. These proteins can be detected in HIV-induced lymphadenopathy as long FDC are present [47, 48]. These findings strongly imply that the role of FDC deserves consideration in the pathogenesis of HIV- and also in FeLV-induced lymphadenopathies.

No attention has been paid to FDC in cats, but extensive studies were carried out in different species. FDC are of nonlymphoid origin and present in primary and secondary follicles. The ultrastructure of FDC in the control

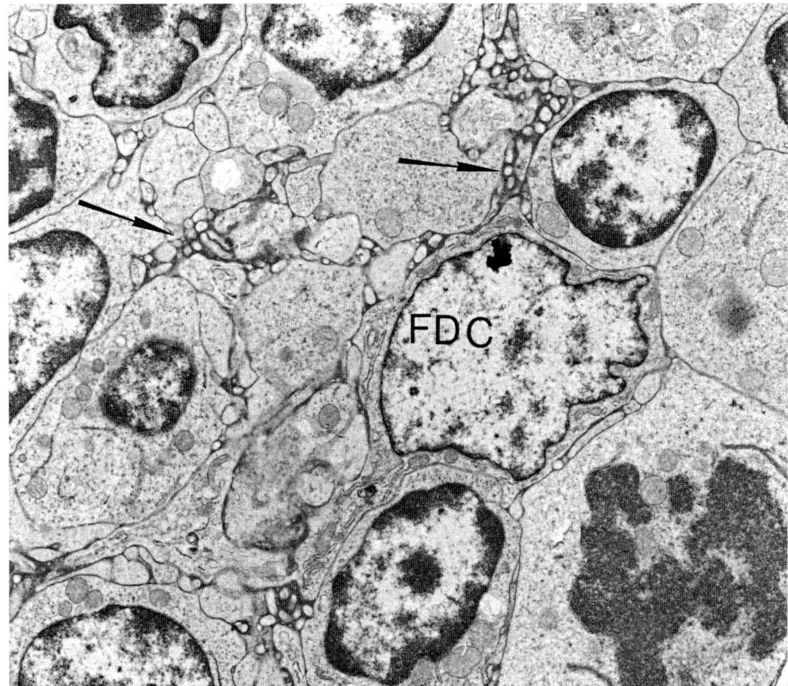

Fig. 4. Part of a germinal center in control cat. The electron micrograph shows a FDC with a narrow rim of cytoplasm around the nucleus, and numerous profiles of dendrites (arrows) covered by electron-dense material. × 2,340.

cat did not differ from those seen in other species. They were large cells containing one or two voluminous, clear nuclear profiles with a thin marginal band of heterochromatin and a distinct nucleolus. Around the nucleus there was a narrow rim of cytoplasm containing only few cell organelles. No ultrastructural signs of phagocytosis could be detected. Characteristically, the cells extended long slender processes (dendrites) in all directions, forming a network (fig. 4). In the meshes of these network centrocytes, centroblasts, macrophages with or without cell phagocytosis and, occasionally, a few lymphocytes were seen. Between the dendrites, desmosomal connections were found (fig. 1). The surface of the cell processes were covered by an electron-dense, amorphous material known to contain antigen-antibody complexes. FDC were present over the entire GC, their density, however, was much higher in the centrocytic zone, facing the marginal sinus, than in the centroblastic one.

The most important function of FDC is to trap and to retain antigens and present them to lymphocytes. Series of experiments have demonstrated that antigens injected in animals accumulate within the GC [for review, see 36]. Autoradiographs of lymphatic tissues from animals who received iodinated Salmonella flagellar antigen have shown that the captured antigen is bound on the cell surface of FDC [28, 29] where it persists for long periods of time [53]. Although FDC are able to bind antigen in absence of specific antibody [10], trapping is highly facilitated if antigen is complexed with antibody. FDC present the antigens to lymphocytes thereby having an important role in the humoral immune response. Tew et al. [53] assume that the persisting antigen could regulate the serum antibody level by a feedback mechanism. Recent data support the notion that presented antigen could stimulate intrafollicular $CD4^+$ lymphocytes [24].

FDC apparently recognize and handle FeLV, and also some other retroviruses, as antigen. Trapping of HIV occurs early after infection and an intrafollicular accumulation of virions can be seen during the course of the disease [51, 52]. Swartzendruber et al. [41] noted C-type virus particles in murine GC. Rauscher leukemia virus has been found in GC of mice as early as 6–24 h after inoculation. At a later time of infection the accumulation of virus particles was evident [11]. The raised amount of retroviruses in GC could be due to continuous entrapment of virions from remote sites of lymph nodes and to intrafollicular virus production.

For HIV-1, both sources have been demonstrated. Budding virions inside the GC have been documented by Armstrong et al. [2, 3] as well as by us [45, 52]. Spread of HIV via lymphatics has also been documented either by using immunohistochemical methods or in situ hybridization [35, 50]. HIV-infected cells, while traveling the marginal sinus, can produce virions. In the present study we have also found infected cells in the marginal sinus which indicate that lymphatics are involved in the spread of FeLV. Budding virions show that the infected cells deliver FeLV directly into the sinus lumen (fig. 5).

Virus particles from the sinus lumen can be directed to the GC. In PGL, due to the exuberant follicular hyperplasia and the loss of mantle zone, dendrites of FDC bulge through the gaps of the sinus wall into the lumen where they can gain immediate contact with HIV particles [50]. Transport of immune complexes from the marginal sinus into the follicles has been demonstrated in animal experiments [21, 42]. Szakal et al. [42] found an active cell-mediated transport of antigen rather than a passive

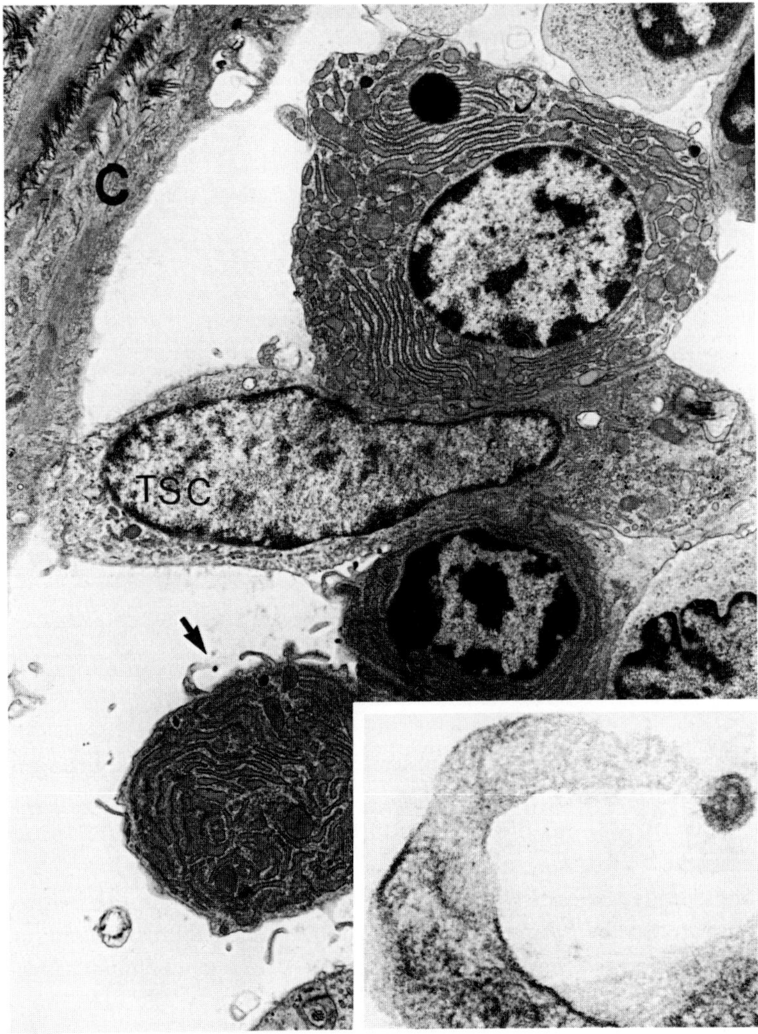

Fig. 5. Marginal sinus with three plasma cells attached to a trabecular sinus endothelial cell (TSC). One of the plasma cells with virus budding (arrow and inset). C = capsule. × 5,450. Inset: × 23,400.

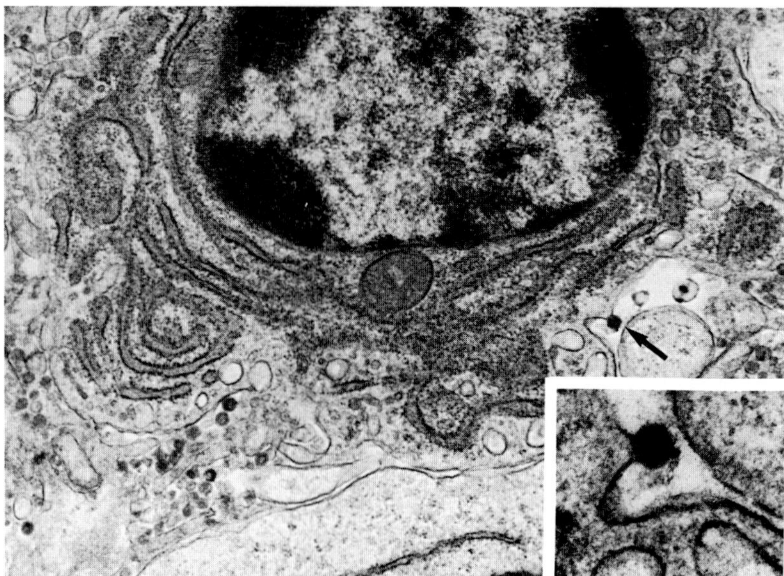

Fig. 6. A plasma cell with virus budding (arrow and inset) inside a germinal center. ×7,800. Inset: ×23,400.

drainage from the sinus into the follicles. A similar pathway can also be postulated for FeLV.

However, the majority of intrafollicular FeLV is probably produced inside the GC. Budding viruses can easily be detected at this location. As in HIV infection, cells of lymphoid and nonlymphoid origin are involved in the intrafollicular virus production. Budding from the cell membrane of plasma cells was frequently seen (fig. 6). Virus-producing plasma cells were found not only in GC but also in the lumen of the marginal sinus (fig. 5). Virions or virus-like particles were not seen in the cytoplasm or in cisternae of the rough endoplasmic reticulum. None of these cells exhibited cellular damage. Centroblasts and large blasts, with abundant rough endoplasmic reticulum which exceeded that of centroblasts, also showed virus budding (fig. 7). These large blasts are probably B immunoblasts, precursors of plasma cells. Thus, these findings demonstrate that cells of the B lineage are very important targets for FeLV. This could be an explanation why Pack and Chapman [31] were not able to detect viruses in any cells of thymuses obtained from FeLV-infected kittens.

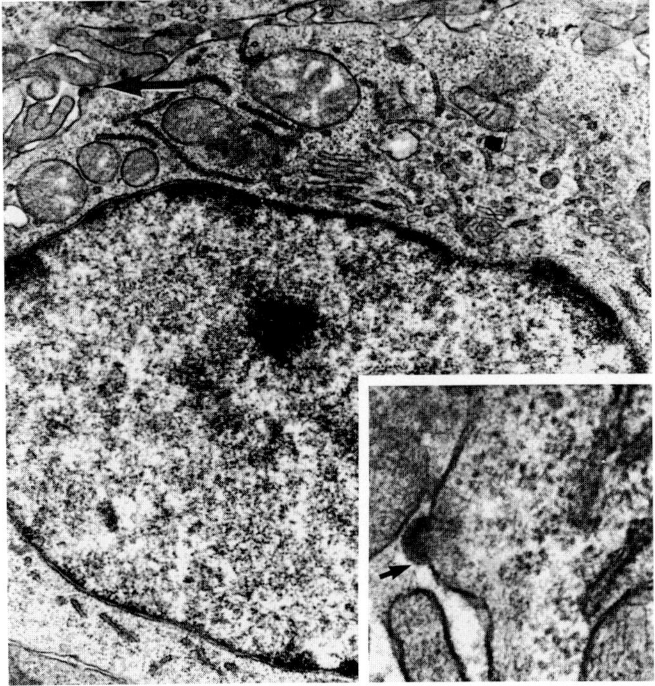

Fig. 7. Budding of FeLV from the cell membrane of a centroblast (arrow and inset). × 7,800. Inset: × 23,400.

This does not correlate with observations made in HIV infection, although it is known that some lymphoblastoid cell lines, obtained by transformation of B lymphocytes by Epstein-Barr virus, can support HIV replication [25]. Findings on an in vivo involvement of plasma cells or their precursors in HIV production have not been documented.

Virus-producing plasma cells have been described in other retrovirus infection. Such plasma cells were found in mice experimentally infected with Friend leukemia virus [22, 23]. In this murine system the infection of plasma cells did not interfere with antibody synthesis. A failure of infected precursor cells to develop into plasma cells has been postulated as a cause of impaired humoral response [22, 23]. On the basis of the present study we cannot decide whether FeLV infect antibody-forming cells at all stages of maturation or only cells at an early stage of differentiation. The relatively higher number of plasma cells with virus budding suggest that even if precursor cells become infected, FeLV does not inhibit the differentiation and maturation of these cells.

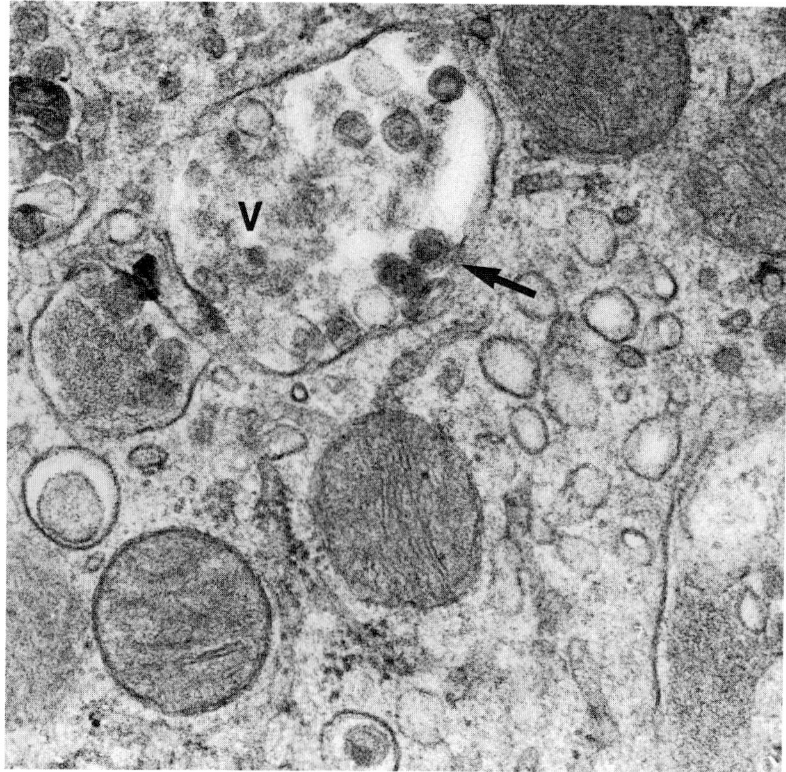

Fig. 8. Several FeLV enclosed in an intracytoplasmic vacuole (V) of a macrophage. FeLV budding into a vacuole (arrow). ×23,400.

The types of nonlymphoid cells involved in intranodal FeLV replication were similar to those in PGL. Macrophages, with FeLV sequestered in intracytoplasmic vacuoles, were frequently seen in GC. Virus budding occurred into the vacuole (fig. 8). The vacuoles contained many virions and the macrophages showed, generally, no ultrastructural signs of cell damage. Free-lying intracellular virions were seen only when macrophages disintegrated.

FDC also support FeLV replication. Budding virions from their cell surface were often observed (fig. 3). We have not found free or vacuole-sequestered virions in the cytoplasm of these cells. We feel that budding from FDC is an especially important finding because it clearly demonstrates that captured FeLV are infectious even if they are complexed with antibody. In contrast to lymphocytes and macrophages, FDC are nonmigratory and long-lived [8]. Therefore, infection of these cells must have occurred inside the GC.

The data, which we gained through the present study, strongly suggest that GC are very important in the pathogenesis of FeLV infection. We assume that the intrafollicular virus accumulation and replication plays an important role in the destruction of GC in FeLV-infected cats. Probably a continuous virus replication occurs here as long as GC exist. Effective antiviral drugs should reach the GC to eliminate these important infectious foci.

References

1 Anderson, L. J.; Jarrett, W. F. H.; Jarrett, O.; Laird, H. M.: Feline leukemia virus infection of kittens: mortality associated with atrophy of the thymus and lymphoid depletion. J. natn. Cancer Inst. *47:* 807–817 (1971).

2 Armstrong, J. A.; Dawkins, R. L.; Horne, R.: Retroviral infection of accessory cells and the immunological paradox in AIDS. Immunol. Today *6:* 121–122 (1985).

3 Armstrong, J. A.; Horne, R.: Follicular dendritic cells and virus-like particles in AIDS-related lymphadenopathy. Lancet *ii:* 370–372 (1984).

4 Baroni, C. D.; Pezzella, F.; Mirolo, M.; Ruco, L. P.; Rossi, G. B.: Immunohistochemical demonstration of p24 HTLV III major core protein in different cell types within lymph nodes from patients with lymphadenopathy syndrome. Histopathology *10:* 5–13 (1986).

5 Biberfeld, P.; Chayt, K. J.; Marselle, L. M.; Biberfeld, G.; Gallo, R. C.; Harper, M. E.: HTLV-III expression in infected lymph nodes and relevance to pathogenesis of lymphadenopathy. Am. J. Pathol. *125:* 436–442 (1986).

6 Diebold, J.; Marche, C. L.; Audouin, J.; Aubert, J. P.; Le Tourneau, A.; Bouton, C. L.; Reynes, M.; Wizniak, J.; Capron, F.; Tricottet, V.: Lymph node modification in patients with the acquired immunodeficiency syndrome (AIDS) or with AIDS-related complex. A histological, immunohistological and ultrastructural study of 45 cases. Pathol. Res. Pract. *180:* 590–611 (1985).

7 Dougherty, E. III; Rickard, C. G.: Ultrastructural studies of feline leukemia virus. J. Ultrastruct. Res. *32:* 472–476 (1970).

8 Fossum, S.; Ford, W. L.: The organization of cell populations within lymph nodes: their origin, life history and functional relationships. Histopathology *9.* 469–499 (1985).

9 Haley, P. J.; Hoover, E. A.; Quackenbush, S. L.; Gasper, P. W.; Macy, D. W.: Influence of antibody infusion of pathogenesis of experimental leukemia virus infection. J. natn. Cancer Inst. *74:* 821–827 (1985).

10 Hanna, M. G.; Francis, M. W.; Peters, L. C.: Localization of ^{125}I-labeled antigen in germinal centers of mouse spleen: effects of competitive injection of specific or non-cross-reacting antigen. Immunology *15:* 75–91 (1968).

11 Hanna, M. G.; Szakal, A. K.; Tyndall, R. L.: Histoproliferative effect of Rauscher leukemia virus on lymphatic tissue: Histological and ultrastructural studies of germinal centers and their relation to leukemogenesis. Cancer Res. Arch. *30:* 1748–1763 (1970).

12 Hardy, W. D., Jr.: Immunopathology induced by the feline leukemia virus. Springer Semin. Immunopathol. *5:* 75–106 (1982).

13 Hardy, W. D.: Feline-acquired immune deficiency syndrome: A feline retrovirus-induced syndrome of pet cats; in Salzman, L. A. (ed.): Animal models of retrovirus infection and their relationship to AIDS, pp. 75–93 (Academic Press, Orlando 1986).

14 Hardy, W. D., Jr.; Essex, M.: FeLV-induced feline-acquired immune deficiency syndrome. Prog. Allergy, vol. 37, pp. 353–376 (Karger, Basel 1986).

15 Hoover, E. A.; Perryman, L. E.; Kociba, G. J.: Early lesions in cats inoculated with feline leukemia virus. Cancer Res. *33:* 145–152 (1973).

16 Hoover, E. A.; Mullins, J. I.; Quackenbush, S. I.; Gasper, P. W.: Pathogenesis of feline retrovirus-induced cytopathic diseases: acquired immune deficiency syndrome and aplastic anemia; in Salzman, L. A. (ed.): Animal models of retrovirus infection and their relationship to AIDS, pp. 59–74 (Academic Press, Orlando 1986).

17 Hoover, E. A.; Zeidner, N. S.; Perigo, N. A.; Quackenbush, S. L.; Strobel, J. A.; Hill, D. L.; Mullins, J. I.: Feline leukemia virus-induced immunodeficiency syndrome in cats as a model for evaluation of antiretroviral therapy. Intervirology *30:* suppl. 1, pp. 12–25 (1989).

18 Hoover, E. A.; Mullins, J. I.; Quackenbush, S. L.; Gasper, P. W.: Experimental transmission and pathogenesis of immunodeficiency syndrome in cats. Blood *70:* 1880–1892 (1987).

19 Jarrett, W. F. H.; Crawford, E. M.; Martin, W. B.; Davie, F.: Leukaemia in the cat. A virus-like particle associated with leukaemia (lymphosarcoma). Nature *202:* 567–568 (1964).

20 Jarrett, W. F. H.; Martin, W. B.; Crighton, G. W.; Dalton, R. G.; Stewart, M. F.: Leukaemia in the cat. Transmission experiments with leukaemia (lymphosarcoma). Nature *202:* 566–567 (1964).

21 Kamperdijk, E. W. A.; Dijsktra, C. D.; Döpp, E. A.: Transport of immune complexes from the subcapsular sinus into the lymph nodes follicles of the rat. Immunobiology *174:* 395–405 (1987).

22 Koo, G. C.; Ceglowski, W. S.; Friedman, H.: Immunosuppression by leukemia viruses. V. Ultrastructural studies of antibody-forming spleens of mice infected with Friend leukemia virus. J. Immun. *106:* 799–814 (1971).

23 Koo, G. C.; Ceglowski, W. S.; Higgins, M.; Friedman, H.: Immunosuppression by leukemia viruses. VI. Ultrastructure of individual antibody-forming cells in the spleens of Friend leukemia virus-infected mice. J. Immun. *106:* 815–830 (1971).

24 Kosco, M.; Szakal, A. K.; Tew, J. G.: In vivo obtained antigen presented by germinal center B cells to T cells in vitro. J. Immun. *140:* 354–360 (1988).

25 Montagnier, L.; Gruest, J.; Chamaret, S.; Dauguet, C.; Axler, C.; Guétard, D.; Nugeyre, M. T.; Barre-Sinoussi, F.; Chermann, J. C.; Brunet, J. B.; Klatzmann, D. D.; Gluckman, J. C.: Adaptation of lymphadenopathy-associated virus (LAV) to replication in EBV-transformed B lymphoblastoid cell lines. Science *225:* 63–66 (1984).

26 deNoronha, F.; Schafer, W.; Essex, M., et al.: Influence of antisera to oncornavirus glycoprotein (gp70) on infections of cats with feline leukemia virus. Virology *85:* 617–621 (1978).

27 deNoronha, F.; Poco, A.; Post, J. E., et al: Virus isolation test for feline leukemia virus. J. natn. Cancer Inst. *58:* 129–131 (1977).

28 Nossal, G. J. V.; Abbot, A.; Mitchell, J.; Lummus, Z.: Antigens in immunity. XV.

Ultrastructural features of antigen capture in primary and secondary lymphoid follicles. J. exp. Med. *127:* 277–289 (1968).

29 Nossal, G. J. V.; Austin, C. M.; Pye, J.; Mitchell, J.: Antigens in immunity. XII. Antigen trapping in the spleen. Int. Archs Allergy appl. Immun. *29:* 368–383 (1966).

30 O'Hara, C. J.; Groopman, J. E.; Federman, M.: The ultrastructural and immunohistochemical demonstration of viral particles in lymph nodes from human immunodeficiency virus-related and non-human immunodeficiency virus-related lymphadenopathy syndromes. Human Pathol. *19:* 545–549 (1988).

31 Pack, F. D.; Chapman, W. L.: Light and electron microscopic evaluation of thymuses from feline leukemia virus-infected kittens. Exp. Pathol. *18:* 95–110 (1980).

32 Parravicini, C. L.; Yago, L.; Costanzi, G. C.: Follicle lysis in lymph node biopsies from patients with persistent generalized lymphadenopathy. Blood *8:* 576–577 (1986).

33 Perryman, I. E.; Hoover, E. A.; Yohn, D. S.: Immunologic reactivity of the cat. Immunosuppression in experimental feline leukemia. J. natn. Cancer Inst. *49:* 1357–1365 (1972).

34 Racz, P.; Tenner-Racz, K.; Kahl, C.; Feller, A. C.; Kern, P.; Dietrich, M.: Spectrum of morphologic changes of lymph nodes from patients with AIDS or AIDS-related complex. Prog. Allergy, vol. 37, pp. 81–181 (Karger, Basel 1986).

35 Racz, P.: Molecular, biologic, immunohistochemical and ultrastructural aspects of lymphatic spread of the human immunodeficiency virus. Lymphology *21:* 28–35 (1988).

36 Racz, P.; Tenner-Racz, K.; Schmidt, H.: Follicular dendritic cells in HIV-induced lymphadenopathy and AIDS. Acta pathol. microbiol. immunol. scand., suppl. *8:* 16–23 (1989).

37 Rickard, C. G.; Post, J. E.; Noronha, F.; Barr, L. M.: A transmissible virus-induced lymphocytic leukemia of cats. J. natn. Cancer Inst. *42:* 987–1014 (1969).

38 Riedel, N.; Hoover, E. A.; Gasper, P. W.; Nicolson, M. O.; Mullins, J. I.: Molecular analysis and pathogenesis of the feline aplastic anemia retrovirus, FeLV-C-Sarma. J. Virol. *60:* 242 (1986).

39 Rojko, J. L.; Hoover, E. A.; Mathes, L. E.; Hause, W. R.; Schaller, J. P.; Olsen, R. G.: Detection of feline leukemia virus in tissues of cats by a paraffin embedding immunofluorescence procedure. J. natn. Cancer Inst. *61:* 1315–1321 (1978).

40 Snyder, H. W., Jr.; Singhal, M. C.; Ernst, N. R.; Grant, C. K.; Cotter, S. M.; Yoshida, L. H.; Jones, F. R.: Extracorporeal perfusion of plasma over immobilized *Staphylococcus aureus* protein A as a treatment for FeLV infection and lymphosarcoma: Prospects for treatment of retroviral infection and AIDS in man; in Salzman, L. A. (ed.): Animal models of retrovirus infection and their relationship to AIDS, pp. 403–419 (Academic Press, Orlando (1986).

41 Swartzendruber, D. C.; Ma, B. I.; Murphy, W. H.: Localization of C-type virus particles in lymphoid germinal centers of C58 mice. Proc. Soc. exp. Biol. Med. *126:* 731–734 (1967).

42 Szakal, A. K.; Holmes, K. L.; Tew, J. G.: Transport of immune complexes from the subcapsular sinus to lymph node follicles on the surface of nonphagocytic cells, including cells with dendritic morphology. J. Immun. *131:* 1714–1727 (1983).

43 Tavares, L.; Roneker, C.; Johnston, K.; deNoronha, F.: 3'-Azido-3'-deoxythymidine in feline leukemia virus-infected cats: A model for therapy and prophylaxis of AIDS. Cancer Res. *47:* 3190–3194 (1987).

44 Tavares, L.; Roneker, C.; Postie, L.; de Noronha, F.: Testing of nucleoside analogues in

cats infected with feline leukemia virus: a model. Intervirology *30:* suppl. 1, pp. 26–35 (1989).

45 Tenner-Racz, K.; Racz, P.; Dietrich, M.; Kern, P.: Altered follicular dendritic cells and virus-like particles in AIDS and AIDS-related lymphadenopathy. Lancet *i.* 105–106 (1985).

46 Tenner-Racz, K.; Racz, P.; Bofill, M.; Schulz-Meyer, A.; Dietrich, M.; Kern, P.; Weber, J.; Pinching, A. J.; Veronese-Dimarzo, F.; Popovic, M.; Klatzmann, D.; Gluckman, J. C.; Janossy, G.: HTLV-III/LAV viral antigens in lymph nodes of homosexual men with persistent generalized lymphadenopathy and AIDS. Am. J. Path. *123:* 9–15 (1986).

47 Tenner-Racz, K.; Racz, P.; Kern, P.; Dietrich, M.: Prognostic and diagnostic value of lymph node biopsy in AIDS and persistent generalized lymphadenopathy; in Staquet, M., Hemmer, R., Baert, H. (eds): Clinical aspects of AIDS and AIDS-related complex, pp. 125–136 (Oxford University Press, Oxford 1986).

48 Tenner-Racz, K.; Racz, P.; Dietrich, M.; Kern, P.; Janossy, G.; Veronese-Dimarzo, F.; Klatzmann, D.; Gluckman, J. C.; Popovic, M.: Monoclonal antibodies to human immunodeficiency virus: their relation to the patterns of lymph node changes in persistent generalized lymphadenopathy and AIDS. AIDS *1:* 95–104 (1987).

49 Tenner-Racz, K.; Racz, P.; Gartner, S.; Dietrich, M.; Popovic, M.: Atypical virus particles in HIV-1-associated persistent generalized lymphadenopathy. Lancet *i:* 774–775 (1988).

50 Tenner-Racz, K.; Racz, P.; Schmidt. H.; Dietrich, M.; Kern, P.; Louie, A.; Gartner, S.; Popovic, M.: Immunohistochemical, electron microscopic and in situ hybridization evidence for the involvement of lymphatics in the spread of HIV-1. AIDS *2:* 299–309 (1988).

51 Tenner-Racz, K.; Racz, P.; Gluckman, J. C.; Popovic, M.: Cell-free HIV in lymph nodes of patients with AIDS and generalized lymphadenopathy. New Engl. J. Med. *318:* 49–50 (1988).

52 Tenner-Racz, K.; Racz, P.; Gartner, S.; Ramsauer, J.; Dietrich, M.; Gluckman, J. C.; Popovic, M.: Ultrastructural analysis of germinal centers in lymph nodes of patients with HIV-1-induced persistent generalized lymphadenopathy: evidence for persistence of infection. Prog. AIDS Pathol. *1:* 29–40 (1989).

53 Tew, J. G.; Phipps, R. P.; Mandel, T. E.: The maintenance and regulation of humoral immune response: persisting antigen and the role of follicular antigen-binding dendritic cells as accessory cells. Immunol. Rev. *53:* 175–201 (1980).

54 Trainin, Z.; Wernicke, D.; Ungar-Waron, H.; Essex, M.: Suppression of the humoral antibody response in natural retrovirus infections. Science *220:* 858–859 (1983).

55 Warner, T. F.; Crass, B.; Gabel, C.; Maki, D.; Hafez, G. D.; Golubjatnikow, R.: Diagnosis of HTLV-III infection by ultrastructural examination of germinal centers in lymph nodes. A case report. AIDS Res. *2:* 43–50 (1986).

Prof. Dr. P. Racz, Department of Pathology and Körber Laboratory for AIDS Research, Bernhard Nocht Institute for Tropical Medicine, Bernhard-Nocht-Strasse 74, D–2000 Hamburg (FRG)

Racz P, Haase AT, Gluckman JC (eds): Modern Pathology of AIDS and Other Retroviral Infections. Basel, Karger, 1990, pp 177–183

HIV Infection of Mononuclear Phagocytes: Role of Antibody-Mediated Virus Binding to Fc-γ Receptors[1]

J. C. Gluckman[a], *T. Jouault*[a], *F. Chapuis*[a], *R. Olivier*[b], *E. Bahraoui*[b], *C. Parravicini*[c]

[a]UER Pitié-Salpêtrière, Paris; [b]Institut Pasteur, Paris, France;
[c]Ospedale L. Sacco, Milano, Italy

It is now recognized that mononuclear phagocytes play an important role in HIV infection: monocytes from patients with AIDS present altered functions such as reduced phagocytosis and chemotaxis [1, 2]; cells that express macrophage markers in different organs, especially in the central nervous system, contain infectious viral particles, viral genome and viral antigens [3, 4]; in vitro, monocyte/macrophage from HIV-infected patients replicate the virus, and monocyte from healthy HIV-seronegative donors are susceptible to HIV infection [5–7]. Thus, beside $CD4^+$ lymphocytes [7, 8], cells of the macrophage lineage are targets for HIV. However, their role in the pathogenesis of AIDS and related syndromes has yet to be more precisely clarified.

Susceptibility of Mononuclear Phagocytes to HIV Infection

The first question that was addressed was whether susceptibility to HIV infection of cells of the macrophage lineage, and the capacity to replicate the virus were related with the degree of cell maturation, as described for other lentiviruses such as ovine visna-maedi and caprine arthritis-encephalitis virus [9–11].

An in vitro culture system to allow blood monocytes to differentiate into macrophages [12] was utilized. Briefly, monocytes from the peripheral blood of normal donors were purified by adherence to plastic before being placed in PTFE

[1]Supported by the European Community, the Körber Stiftung and the PNRS.

vials for long-term culture (≥4 weeks) in culture medium (RPMI 1640 enriched with 50% pooled normal human serum) in the absence of any additional growth factor. Under these conditions, normal monocytes differentiated into mature macrophages within 4–7 days of culture: beside characteristic morphological changes, peroxidase staining decreased while sodium fluoride inhibitable alpha-naphtyl butyrate esterase and ASD acetate esterase activity increased; in parallel, membrane expression of differentiation markers CD16 (FcRγIII), CD35 (CR1) and CD11b (CR3) increased, while CDW32 (FcRII) and CD14 expression remained stable; expression of the CD4 molecule (the cellular receptor for HIV [8]), which was limited and almost undetectable at the monocyte membrane, increased approximately by 8-fold during this period and remained stable thereafter. The capacity of the cells to bind HIV particles, as determined with anti-gp110 murine monoclonal antibodies (mAb) according to the technique described by McDougal et al. [13], augmented to a similar degree.

This observation led to investigate whether susceptibility to HIV infection increased in parallel during maturation from monocytes to macrophages. The ability of monocytes, macrophages taken either after 3 or 7 days of culture, and lymphocytes deprived of monocytes to transmit HIV infection to recipient cultures was compared [14]. Cell preparations were inoculated with HIV-LAV_{BRU} for 18 h and washed free of excess virus. Thereafter, various cell numbers (from 1 to 10^5) were added to 8 replicates of 1×10^5 PHA-stimulated lymphocytes in microplates, and the latter cultures were monitored for virus replication by the HIV antigen capture assay [15]. The percentage of cultures that were positive after 7 days was then determined. Between 1×10^1 and 2×10^2 HIV pulsed monocytes or macrophages were required to elicit productive infection in 50% of recipient PHA-blast cultures, as compared with $1–2 \times 10^3$ cells from monocyte-deprived lymphocyte preparations. Although the cell dose required for transmission of infection in this assay was a function of both the number of cells to the membrane of which viral particles could have simply adsorbed and the frequency of cells that were actually infected, and not strictly the latter, these results strongly suggest that there is no significant difference in the frequency of infected monocytes or macrophages under the conditions used. However, these cell types appear to be at least 10-fold more susceptible to HIV than lymphocytes, which contrasts with what others have reported when using different experimental procedures [15].

In another series of experiments, the time course of HIV production by fresh monocytes and 5-day differentiated macrophages following exposure to HIV-1 LAV_{BRU} was examined. Peak viral production, as determined by the measure of HIV antigen and reverse transcriptase activity in the culture supernatants, was attained later (by approximately 3 days) in monocytes than in macrophages. This difference of the kinetics of viral replication could be related either to differences in the frequency of initially infected cells, or to restricted replication, in monocytes versus macrophages. The results described above rule out the first possibility. Therefore, it appears probable that undifferentiated monocytes are permissive to HIV but cannot efficiently replicate the virus, while differentiation into macrophages is accompanied by increased viral replication. The fact that the level of HIV replication may be differentially regulated at distinct stages of mononuclear phagocyte maturation is very similar to the pattern of visna virus replication in normal sheep monocytes and macrophages [10, 11, 16]. The state of monocyte differentiation and its role on virus replication is certainly an important component in HIV infection and in the pathogenesis of AIDS [3, 7, 11].

Role of Antibody-Mediated HIV Binding to Fc-γ Receptors in the Infection of Mononuclear Phagocytes

Surprisingly, in the course of these experiments, HIV infection of either monocytes or macrophages was not inhibited when virus had been preincubated with a mouse mAb directed to gp110 and known to neutralize infection of $CD4^+$ lymphoid cells (mAb 110-4 from Genetic Systems) [17–19]. Because this mAb was of the IgG1 isotype, which has been reported to be able to bind macrophages' FcR-I and FcR-II [20], we hypothesized that as for other viruses (such as flaviviruses [21, 22] or lentiviruses [23, 24]) IgG directed against viral envelope glycoproteins may play a role in the infection of cells expressing Fc-γ receptors (FcR) for IgG.

This led us to investigate whether human IgG directed to gp110 may serve as an attachment system to FcR, allowing eventual infection of mononuclear phagocytes. An anti-HIV preparation that comprised antibodies that recognized HIV gp110 (HIV-IgG) was selected. When analyzed by flow cytometry, these HIV-IgG specifically labeled virion-coated CEM cells, and they prevented viral particles from binding to CD4 at the cell membrane when virus was first preincubated with HIV-IgG before adding the cells [13]. Similar results were obtained when using soluble vaccinia-virus-recombinant radiolabeled *gp160. In addition, HIV-IgG strongly inhibited HIV infection of CEM cells: as compared with control cultures, viral replication was significantly delayed when the virus had been preincubated with HIV-IgG before infection. Thus, the HIV-IgG preparation used comprised antibodies that recognized HIV gp110 and that prevented virions from binding to CD4 blocking virus attachment to, and infection of, the lymphoid cells.

Quite different results were obtained when the human monocytic cell line, U937, was used as a model for HIV-monocyte interactions. U937 cells are susceptible to HIV [25] and are known to express both low and high affinity FcR in addition to CD4 [26–29]. First, HIV-IgG could be detected at the cell membrane, whether or not viral particles had been added, and no modification of binding was noted whether virus had been preincubated or not with HIV-IgG; however, when $F(ab')_2$ fragments were used instead of complete HIF-IgG, no labeling of the cells was noted anymore in the absence of virus, while fluorescence intensity increased in a dose-dependent manner when the virus had first adsorbed to U937 cells, and total inhibition of binding occurred when virus had been preincubated with HIV-IgG $F(ab')_2$. These results indicate that HIV-IgG that specifically recognize gp110 can bind to membrane FcR and elicit HIV to attach to monocytic cells. Second, precoating U937 cells with HIV-IgG or complexing the antibodies with soluble *gp160 resulted in increased fixation of *gp160 to the cells, which was inhibited by heat-aggregated normal human IgG [30]. Therefore, beside binding to CD4, HIV or soluble envelope glycoprotein can in addition attach to FcR-bearing cells by way of specific antibodies.

It was then examined whether this IgG-dependent mode of attachment could actually result in infection of the monocytic cells by HIV. In contrast with what was noted with CEM cells, the kinetics and intensity of HIV replication in U937 cells was similar whether these cells had been infected with free or with immune-complexed viral particles. This observation indicates that, at least, there is no significant reduction of the number of cells initially infected in the presence of HIV-IgG and that the fixation of immune-complexed HIV to FcR may lead to infection of cells of the macrophage lineage as already shown for other viruses [21–24]. However, enhancement of infectivity, such as what others have described with other viruses under comparable conditions [22, 31], or with HIV in a different experimental model [32], was only limited even when HIV-IgG were titrated down to balance their possible inhibiting capacities [21, 24]. In addition, it appeared in the same experiments that, even when antibodies served as an attachment system for HIV, interaction of gp110 with CD4 was nonetheless still necessary for this route of infection, since treatment of the cells with the anti-CD4 mAb Leu3a, known to interfere with the CD4 site that interacts with gp110 [33], resulted in strong inhibition of viral production whether or not virus had been immune-complexed with HIV-IgG.

Thus, antibodies directed to gp110, which hinder the interaction with CD4 without necessarily recognizing the site involved in the binding to CD4 [34], normally rather prevent contact between immune-complexed HIV and the cell membrane. In the case of mononuclear phagocytes that possess FcR, these antibodies may, on the contrary, serve as an attachment system for HIV particles. A particle dissociating from a membrane FcR would then have greater probability of encountering a CD4 molecule at the cell surface than it has of diffusing into the medium above the surface [35]. Therefore, such attachment might then eventually allow contact between viral gp110 and the CD4 receptor, and result in productive infection of the cells. On the other hand, it is also possible that FcR binding leads to cellular infection via the internalization of viral particles into phagosomes [36] without any interaction with CD4, which nonetheless would then be required for further rounds of replication of viral progeny.

Taken together, these results indicate that antibodies directed at gp110 may play a role for allowing HIV infection of FcR-bearing cells. This may well represent a mechanism by which HIV may spread in vivo in the macrophage 'reservoir', and even circumvent the possible neutralizing effect of antibodies.

References

1 Poli G, Bottazzi B, Acero R, et al: Monocyte function in intravenous drug abusers with lymphadenopathy syndrome and in patients with acquired immunodeficiency syndrome: selective impairment of chemotaxis. Clin Exp Immunol 1985;62:136–142.
2 Estevez ME, Ballart IJ, Diez RA, et al: Early defect of phagocytic cell function in subjects at risk for acquired immunodeficiency syndrome. Scand J Immunol 1986;24:215–221.
3 Ho DD, Pomerantz R, Kaplan J: Pathogenesis of infection with human immunodeficiency virus. N Engl J Med 1987;317:278–286.
4 Price RW, Brew B, Sidtis J, et al: The brain in AIDS: Central nervous system HIV-1 infection and AIDS dementia complex. Science 1988;239:586–592.
5 Gartner S, Markovits P, Markovitz DM, et al: The role of mononuclear phagocytes in HTLV III/LAV infection. Science 1986;233:215–219.
6 Gendelman HE, Orenstein JM, Martin MA, et al: Efficient isolation and propagation of human immunodeficiency virus on recombinant colony-stimulating factor 1-treated monocytes. J Exp Med 1988;167:1428–1441.
7 Fauci AC: the human immunodeficiency virus: infectivity and mechanisms of pathogenesis. Science 1988;239:617–622.
8 Klatzmann D, Gluckman JC: HIV infection: facts and hypotheses. Immunol Today 1986;7:291–296.
9 Narayan O, Kennedy-Stoskopf D, Sheffer DE, et al: Activation of caprine-arthritis-

encephalitis virus expression during maturation of monocytes to macrophage. Infect Immunol 1983;41:67–73.
10 Gendelman HE, Narayan O, Kennedy-Stoskopf D, et al: Tropism of sheep lentiviruses for monocytes: susceptibility to infection and virus gene expression increasing during maturation of monocytes to macrophages. J Virol 1986;58:67–74.
11 Haase AT: Pathogenesis of lentivirus infections. Nature 1986;322:130–136.
12 Crowe S, Mills J, McGrath MS, et al: Quantitative immunocytofluorographic analysis of CD4 surface antigen expression and HIV infection of human peripheral blood monocyte/macrophages. AIDS Res 1987;3:135–145.
13 McDougal JS, Mawle A, Cort SP, et al: Cellular tropism of the human retrovirus HTLV-III/LAV.1. Role of T cell activation and expression of the T4 antigen. J Immunol 1985;135:3151–3162.
14 Nicholson JKA, Cross GD, Callaway CS, et al: In vitro infection of human monocytes with human T lymphotropic virus type III/lymphadenopathy-associated virus (HTLV-III/LAV). J Immunol 1986;137:323–329.
15 McDougal, Cort SP, Kennedy MS, et al: Immunoassay for the detection and quantitation of infectious human retrovirus, lymphadenopathy-associated virus (LAV). J Immunol Methods 1985;76:171–183.
16 Narayan O, Sheffer D, Clements JE, et al: Restricted replication of lentiviruses: Visna viruses induce a unique interferon during interaction between lymphocytes and macrophages. J Exp Med 1985;165:1954–1969.
17 Bahraoui EM, Clerget-Raslain B, Chapuis F, et al: A molecular mechanism of inhibition of HIV-1 binding to $CD4^+$ cells by monoclonal antibodies to gp110. AIDS 1988;2:165–169.
18 Kinney-Thomas E, Weber JN, McClure J, et al: Neutralising monoclonal antibodies to the AIDS virus. AIDS 1988;2:25–29.
19 Linsley PS, Ledbetter JA, Kinney-Thomas E, et al: Effect of anti-gp120 monoclonal antibodies on CD4 receptor binding by the *env* protein of human immunodeficiency virus type I. J Virol 1988;62:3695–3702.
20 Unkeless JC, Scigliano E, Freedman V: Structure and function of human and murine receptors for IgG. Annu Rev Immunol 1988;6:251–281.
21 Peiris JSM, Porterfield JS: Antibody mediated enhancement of flavivirus replication in macrophage-like cell lines. Nature 1979;282:509–511.
22 Gollins SW, Porterfield JS: Flavivirus infection enhancement in macrophages: radioactive and biological studies on the effect of antibody on viral fate. J Gen Virol 1984;65:1261–1272.
23 Legrain P, Goud G, Buttin G: Increase of retroviral infection in vitro by the binding of antiviral antibodies. J Virol 1986;60:1141–1143.
24 Kennedy-Stoskopf S, Narayan O.: Neutralizing antibodies to visna lentivirus: mechanism of action and possible role in virus persistence. J Virol 1986;59:37–44.
25 Asjö B, Ivhed I, Gillund M, et al: Susceptibility to infection by the human immunodeficiency virus (HIV) correlates with T4 expression in a parental monocytoid cell line and its subclones. Virology 1987;157:359–365.
26 Wood GS, Warner NL, Warnke RA: Anti-Leu3/T4 antibodies react with cells of monocyte/macrophage and Langerhans lineage. J Immunol 1983;131:212–216.
27 Stewart SJ, Fujimoto J, Levy R: Human T lymphocytes and monocytes bear the same Leu-3 (T4) antigen. J Immunol 1986;136:3773–3778.

28 Jones DH, Looney RJ, Anderson CL: Two distinct classes of IgG Fc receptors on a human monocyte line (U937) defined by differences in binding of murine IgG subclasses at low ionic strength. J Immunol 1985;135:3348–3353.
29 Looney RJ, Abraham GN, Anderson CL: Human monocytes and U937 cells bear two distinct Fc receptors for IgG. J Immunol 1986;136:1641–1647.
30 Yagawa K, Onoue K, Aida Y: Structural studies of Fc receptors. I. Binding properties, solubilization and partial characterization of Fc receptors of macrophages. J Immunol 1979;122:366–373.
31 Barett ADT, Gould EA: Antibody-mediated early death in vivo after infection with yellow fever virus. J Gen Virol 1986;67:2539–2542.
32 Robinson WE, Montefiori DC, Mitchell WM: Antibody-dependent enhancement of human immunodeficiency virus type I infection. Lancet 1988;i:790–794.
33 Sattentau QJ, Dalgleish AG, Weiss RA, et al: Epitopes of the CD4 antigen and HIV infection. Science 1986;234:1120–1123.
34 Looney DJ, Fisher AG, Putney SD, et al: Type-restricted neutralization of molecular clones of human immunodeficiency virus. Science 1988;241:357–359.
35 Edidin M: Function by association? MHC antigens and membrane receptor complexes. Immunol Today 1988;9:218–219.
36 Pauza CD, Price TM: Human immunodeficiency virus infection of T cells and monocytes proceeds via receptor-mediated endocytosis. J Cell Biol 1988;107:959–968.

Jean-Claude Gluckman, MD, CERVI, Hôpital de la Pitié-Salpêtrière,
83, Bd. de l'Hôpital, F–75634 Paris Cédex 13 (France)

Racz P, Haase AT, Gluckman JC (eds): Modern Pathology of AIDS and Other Retroviral Infections. Basel, Karger, 1990, pp 184–200

Immunohistochemical Study of Lymphoid Tissue in HIV-Infected Patients

J. Audouin[a], *G. Szekeres*[a], *A. Le Tourneau*[a], *S. Prevot*[a], *J. G. Fournier*[b], *J. Diebold*[a,1]

[a]Service Central d'Anatomie et de Cytologie Pathologiques, Hôtel-Dieu, et [b]INSERM, U. 43, Hôpital Saint Vincent de Paul, Paris, France

Immunohistochemical studies have made it possible to analyze the cell populations which constitute the lymph node and splenic tissue in HIV-infected patients. These have also permitted the localization of viral proteins in certain areas.

Methods and Materials

Tissue frozen in liquid nitrogen and stored at –80°C was used to determine the phenotypes of T and B lymphocytes, histiocytes and accessory cells of the immunity and to detect viral antigens. Several monoclonal antibodies marked with peroxidase were used according to an indirect technique. Amplification was performed with the ABC labelling procedure and the reaction revealed with aminoethylcarbazole as a chromogen. Intracytoplasmic immunoglobulins were identified on paraffin sections after fixation in Bouin's solution or formol using polyclonal antibodies according to a direct method with fluorescein isothiocyanate or peroxidase as a marker.

Study of Cell Populations of Lymphoid Tissue

This study chiefly involves the examination of the lymph nodal and splenic tissue at different phases in the evolution of the infection. Histological modifications of lymph nodes were assessed according to a previously presented classification [10, 15]. This classification distinguishes a type IA character-

[1]We are grateful to Mr. Wolfelsperger for providing the photographs.

Table 1. Comparison between the European classification and ours

Our classification	European classification
	Basic features
Type IA	Follicular hyperplasia without (FH–FF) or with (FH + FF) severe fragmentation
Type II	Follicular involution (Fi)
Type III	Diffuse pattern (D)
	Special features
	Angioimmunoblastic hyperplasia AIH
Type IB	Multicentric Castleman-like lesions
Type IVA od B	Vascular lesions (pretumorous and tumorous stage of Kaposi's sarcoma)
	Other features

ized by a predominant follicular hyperplasia (FH with or without fragmentation in European classification [20]), a type II characterized by a diffuse, angioimmunoblastic-like appearance (Fi in European classification) with or without angioimmunoblastic-like features, a type III with lymphoid depletion (D in European classification) and a type IB resembling multicentric Castleman's syndrome. Table 1 presents the correlation between the European classification [20] and our own [10, 15].

The histological modifications of splenic lymphoid tissue were assessed according to a classification presented in a recent study [27]. In the spleens removed from HIV-positive patients with thrombopenic purpura, PGL, ARC or early AIDS symptoms, the follicles of the white pulp were sometimes hyperplastic with large germinal centres (IA), sometimes resembled Castleman's syndrome with centrofollicular depletion (IB), were sometimes involuted without Castleman's features (IC) and sometimes atrophic (III). These features can coexist in the same spleen. The predominance of one particular feature makes it possible to distinguish different forms called IA, IB, IC, III.

T Lymphocyte Populations

In the lymph nodes with type IA pattern, the total number of T lymphocytes was initially normal or only slightly decreased in comparison with control

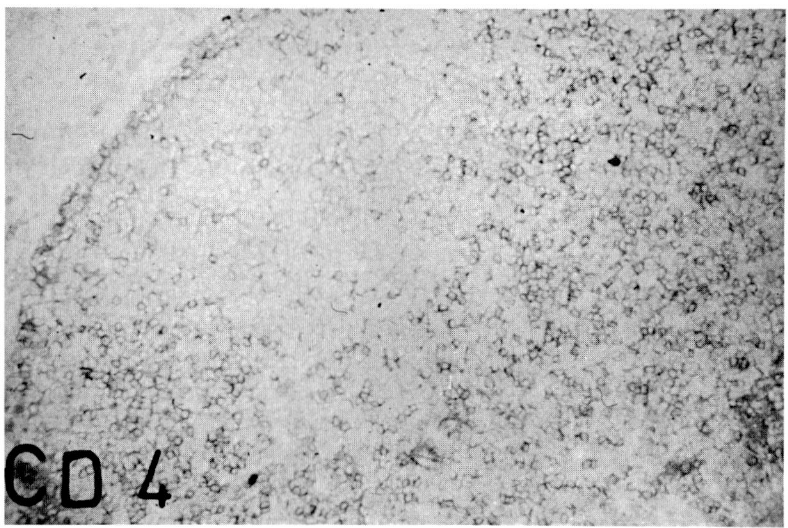

Fig. 1. Lymph node IA type (FH without FF). Only a few CD4-positive lymphocytes are present in the germinal center, but many are seen in the deep cortical area.

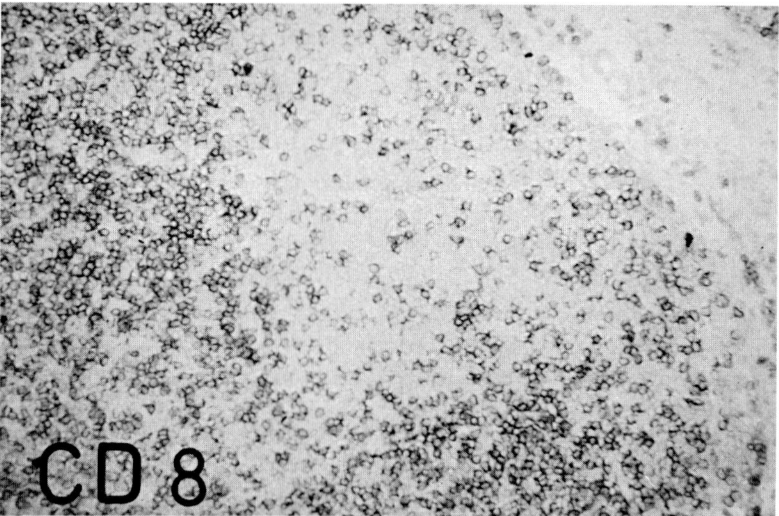

Fig. 2. Same lymph node. Many CD8-positive lymphocytes are present in and around a germinal center.

cases [9]. The CD4-positive lymphocytes were present in a more or less decreased number in the deep cortex, in the interfollicular area and in the germinal centers [2, 3, 5, 7, 9–12, 16, 17, 21, 22, 24–26, 28, 33] (fig. 1). The number of CD8-positive lymphocytes was increased in the same regions (fig. 2). The most important finding was the increase in CD8-positive cells in the germinal centers [2, 3, 5, 7–13, 16, 21, 22, 24–26, 28, 33, 34, 36]. This was a constant finding in HIV-infected patients and, by contrast, was virtually absent or rarely seen in control cases [8]. So this finding can be regarded as highly suggestive of an HIV infection. These CD8-positive lymphocytes were present in the germinal centers as clusters or dispersed cells [2, 3, 5, 7–13, 16, 21, 22, 25, 26, 28, 33, 34, 36]. CD8-positive lymphocytes can also be found in the sinuses [32].

As the disease progressed, the number of CD4 lymphocytes decreased in types II and III. The CD8 lymphocytes were predominant in both types [2, 3, 5, 7, 10–13, 16, 17, 21, 22].

In the spleen with type IA pattern [27], the same modifications were observed in the follicles (fig. 3, 4) and in the periarterial lymphoid sheets (fig. 5). In types IC and III, the number of CD4-positive lymphoid cells in these areas was dramatically reduced and the quantity of CD8-positive cells was normal or increased. These latter cell types were numerous in the red pulp cords in spleens with type IA, IB and IC pattern.

B Lymphocyte Populations

In lymph node with the IA type, the B hyperplasia was characterized by well-developed germinal centers consisting of a polyclonal centrofollicular population with a predominant heavy mu chain. Identification of the membrane IgD immunoglobulin showed an atrophy of the perifollicular mantle zone which varied in size (fig. 6). Occasionally this mantle zone disappeared completely on all or part of the circumference of the germinal centers [2, 10–12, 16, 17, 21, 22, 24]. IgD-positive lymphocytes were sometimes present in the lymphoid infiltrates of the germinal centers. A considerable number of IgD cells were scattered in the interfollicular areas and in the deep cortex.

The B lymphoid hyperplasia was also characterized by a marked polyclonal plasmacytosis (medullary cords, germinative centers, external cortex, occasionally deep cortex). These plasmocytes secrete the three principal immunoglobulins. In most cases they chiefly produce IgG [2]. In some cases, the secretion of IgA is predominant [2, 10–12].

The B markers also revealed the presence of a varying number of immunoblasts in the deep cortical and interfollicular areas [2, 10–12]. In most cases

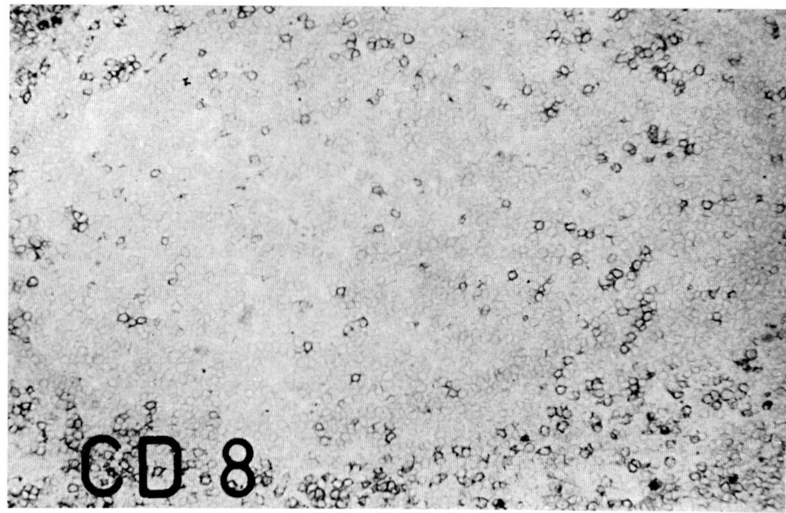

Fig. 3. Spleen, IA type. The CD8-positive lymphocytes are slightly increased in the germinal center of a follicle.

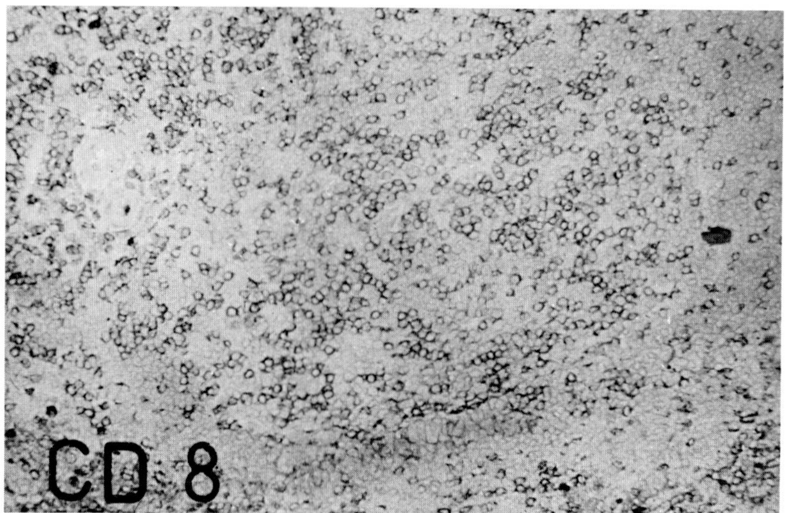

Fig. 4. Spleen, IA type. A very high number of CD8-positive cells can be recognized in another follicle.

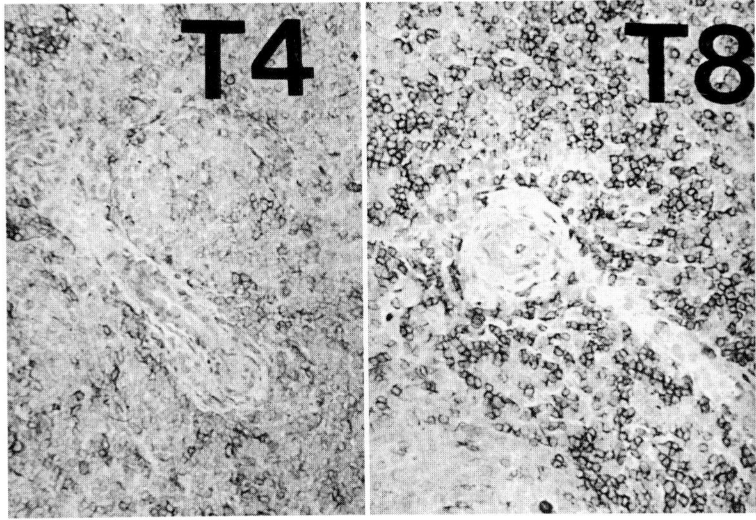

Fig. 5. Spleen, IC type. Severe depletion of the periarteriolar sheet in CD4-positive cells contrasting with an increased number of CD8-positive lymphocytes.

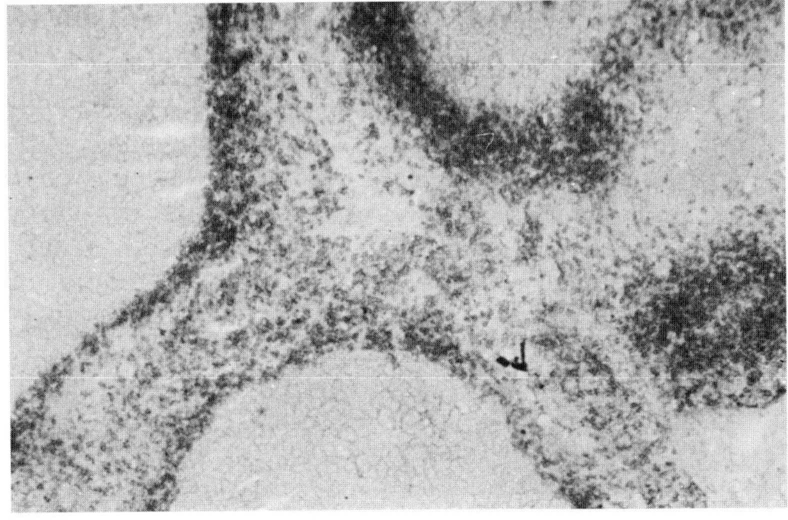

Fig. 6. Lymph node, IA type (FH without FF). Notice the more or less advanced atrophy of the IgD-positive mantle zone of three follicles.

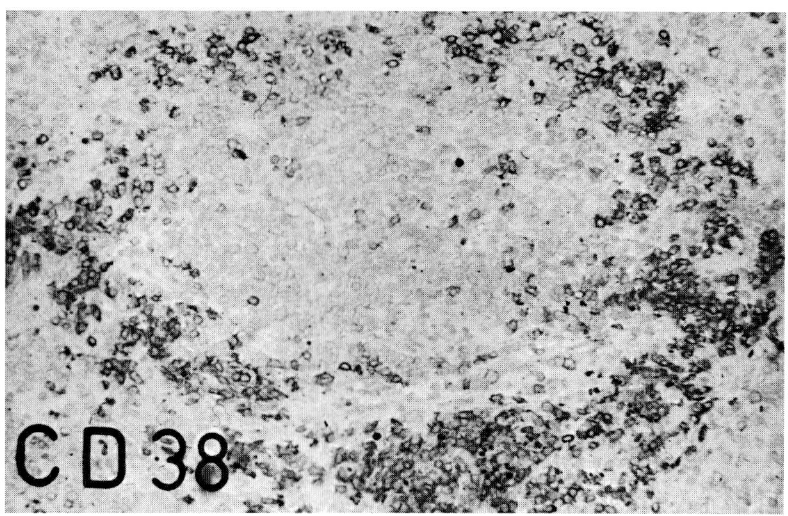

Fig. 7. Spleen, IB type. Presence of a high number of B cells in plasma cell differentiation as shown by anti CD38 monoclonal antibody (OKT 10).

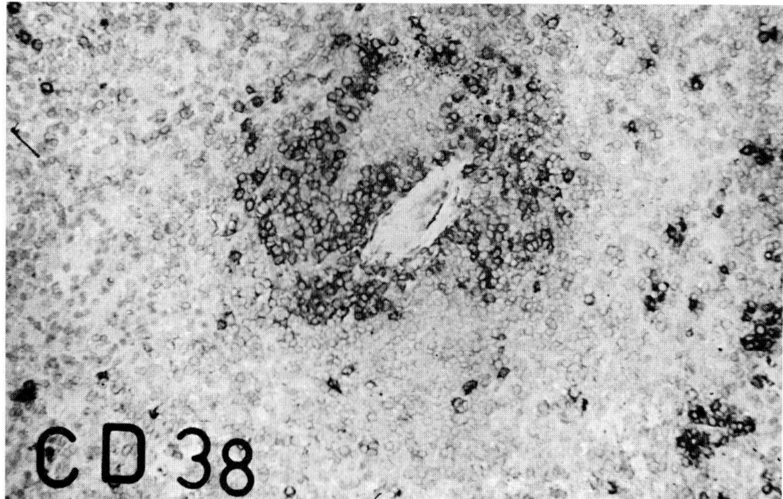

Fig. 8. Spleen. IC type. Most of the cells which constitute the periarteriolar lymphoid sheet are in fact plasma cells or precursors of plasma cells.

they also detected foci of sinusal B lymphocytosis expressing a membrane immunoglobulin which was either mu or gamma [2, 10–12].

B hyperplasia, especially mu and gamma polyclonal plasmocyte hyperplasia, was present in type IB lymph node as in all cases of multicentric Castleman's syndrome. In the follicles, the B cell markers also revealed a depletion in centrofollicular cells and the presence of plasmocytes.

In type II lymph node, the immunoblasts and polyclonal plasmocytes remained numerous though they did diminish in number. This reduction was even more marked in type III. However, despite this, in forms with severe depletion, the B cells nevertheless represented the majority of lymphoid cells, especially in the form of plasmocytes and immunoblasts. As certain infections progressed, large areas of polyclonal plasmocytes could be seen in the cortex and medulla.

In the spleen [27], the B hyperplasia was characterized essentially by a significant polyclonal plasmacytosis secreting mainly the gamma and sometimes the mu heavy chain, and only rarely the alpha heavy chain. This plasmacytosis was particularly marked in spleens with lymphoid atrophy (type IC and III). The hyperplasia extended to areas normally occupied by the plasmocytes, that is the cords of the red pulp around the terminal arterioles. It was associated with a severe plasmacytosis of unusual topography: marginal zone of follicles (fig. 7) and periarteriolar lymphoid sheets (fig. 8). In spleens with lymphoid depletion (type IC and III), the majority of cells in the periarteriolar lymphoid sheets were plasmacytes. Finally, in patients with multiple infections, a severe pseudotumorous plasmacytosis could be observed in the red pulp.

Type IA and IB hyperplastic follicles were polyclonal, with no unusual features. However, the frequent integrity of the perifollicular mantle zones must be stressed, demonstrated by the identification of lymphocytes expressing a membrane IgD immunoglobulin. This mantle zone can be more or less normal even though, in the same patient, a peripheral lymph node may show marked atrophy of the mantle zone. It seems that modifications of the mantle zone appear later and are less severe in the spleen [27] than in the lymph nodes (fig. 9).

Histiomonocyte Populations

Immunostaining studies provided little significant information. They confirmed the presence of a histiocyte hyperplasia in the lymph nodes at all stages, in the form of sinusal histiocytosis and/or histiocyte or epithelioid granulomas. This hyperplasia seemed to be less marked during lymphoid

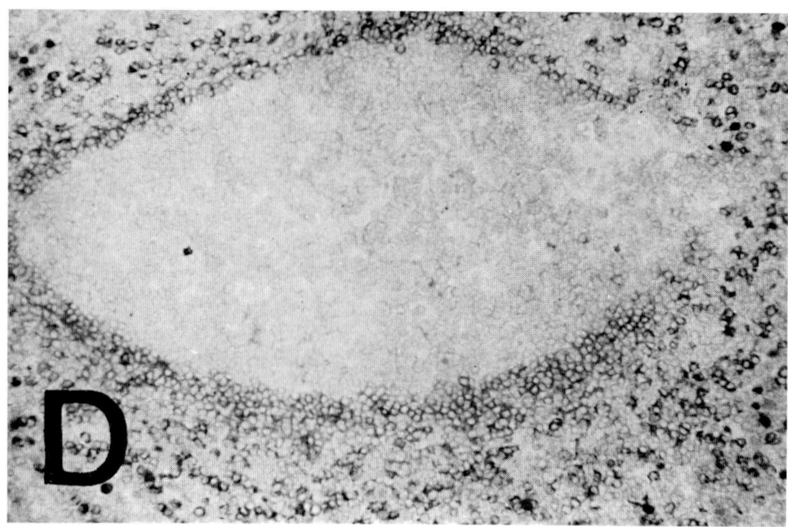

Fig. 9. Spleen, IA type. The mantle zone of this follicle as atrophic is shown with an anti-IgD monoclonal antibody.

depletions of type III. It must be noted here that in the sinus, the histiocytes were associated with CD1 and S 100 protein-positive interdigitating and/or Langerhans' cells [2, 10–12, 31, 32]. These elements were accompanied by T lymphocytes expressing most often CD8 and occasionally CD4 [32] and by polyclonal plasmocytes. In the spleens, numerous histiocytes, sometimes filled with hemosiderin, were present in the cords as well as histiocytic or epithelioid granulomas.

Accessory Cells of the Immunity

Follicular Dendritic Cells. These cells can be identified either by using specific monoclonal antibodies (DRC 1), or by antibodies which recognize the receptor of the C3b or CR 1 (IOB 1A, CD 35). In the germinal center they formed a network of tight mesh between the centrofollicular B cells. In the mantle zone, they were disposed in several concentric layers [2, 10–12].

In type IA lymph nodes, more or less extended areas of the germinal centers did not show any staining of follicular dendritic cells. These negative zones had irregular outlines, like a map (fig. 10). This disruption grew pro-

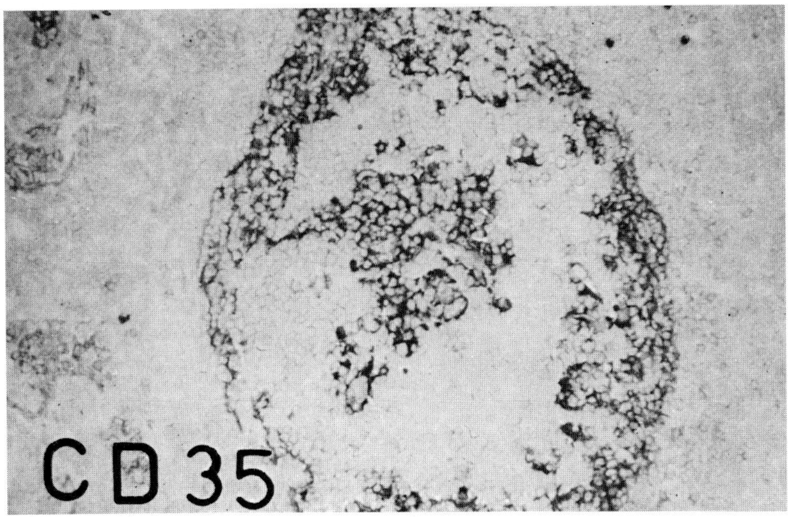

Fig. 10. Lymph node, IA type (FH with FF). The follicular dendritic cell network is disrupted. Irregular areas are unstained with anti-CD35 monoclonal antibody, recognizing the C3b receptor (RC1).

gressively to the point where complete lysis of the germinal centers occurred. The progressive fragmentation of the network of follicular dendritic cells represented the largest and most significant lymph node lesion [2, 3, 5, 7, 10–13, 16–19, 21, 22, 24–26, 34, 36]. The ultrastructural study confirmed the existence of cytoplasmic alterations of the extensions of these cells [10, 14, 29].

In type IB lymph nodes, the dendritic reticular cells were arranged in light concentric circles because of the collapse caused by centrofollicular B cell depletion.

In the lymph nodes with relatively advanced lymphoid depletion (type II and III), the staining made it possible to detect some dispersed clusters of follicular dendritic cells, remnants of the dislocated follicles.

In the splenic parenchyma [27], alterations in the network of follicular dendritic cells were less marked, and sometimes absent altogether in spleens with type IA appearance (fig. 11). In the type IB follicles, the same concentric arrangement of follicular dendritic cells could be observed. In more advanced stages of lymphoid depletion the network is atrophic and more or less broken up (fig. 12).

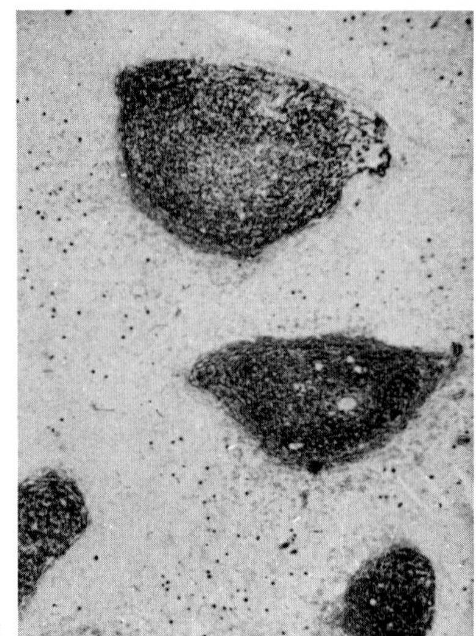

11

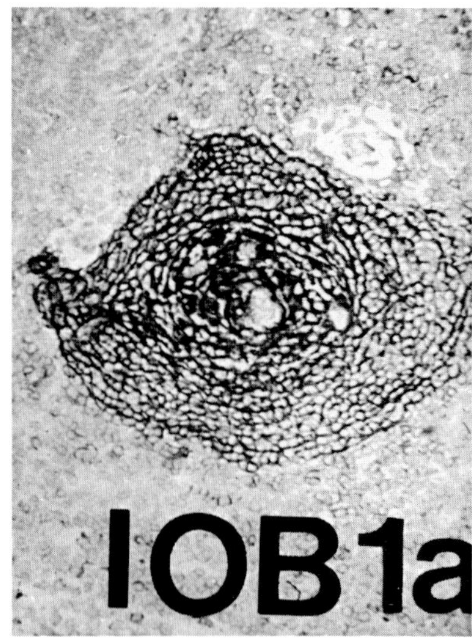

12

Fig. 11. Spleen, IA type. Normal follicular dendritic network demonstrated with a monoclonal antibody recognizing the C3b receptor (IOB1A, CD35).

Fig. 12. Spleen, IB type. Notice the onion bulb pattern of the follicular dendritic cell in this follicle (IOB1A or CD35).

Interdigitating Cells. These cells were identified by a monoclonal antibody anti-CD1 (T6) on frozen tissue and by the polyclonal antibody recognizing the protein S100 on paraffin sections, and showed little change. In the type IA and IB lymph nodes, these were sometimes more numerous than usual. Most often they were present in apparently normal quantities. Their number was normal or slightly reduced in relatively advanced stages of lymphoid depletion. Some interdigitating cells were disclosed in the lymph node sinuses.

Identification of Viral Proteins

Monoclonal antibodies directed against the protein p24 of the viral core (kindly provided by G. Pallesen, Aarhus, Denmark, and Mandrand, Lyon, France) enabled identification of this viral protein chiefly in germinal centers. This consisted of thick deposits of positivity between the centrofollicular lymphoid cells (fig. 13, 14). Staining revealed a network similar to that of

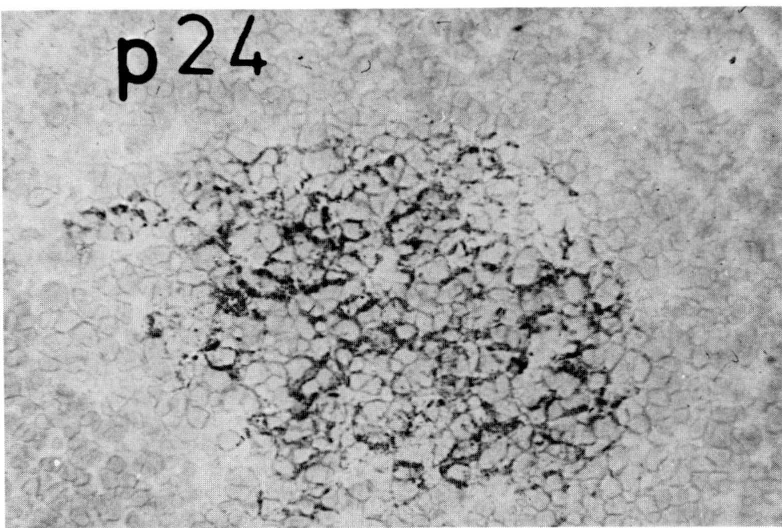

Fig. 13. Lymph node IA type (FH with FF). The positivity for the HIV p24 protein is seen in the germinal center, realizing a pattern similar to the labelling of the follicular dendritic cells.

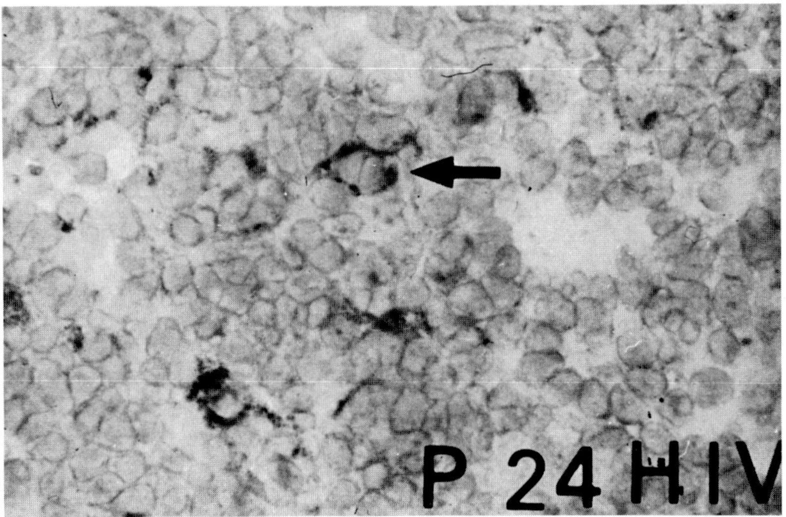

Fig. 14. Spleen. IA type. The positivity for HIV p24 protein is seen surrounding the two nuclei of a follicular dendritic cell. This pattern is suggestive of a cytoplasmic positivity.

follicular dendritic cells obtained with specific antibodies. Certain aspects, especially the stained zones around the double nuclei of the follicular dendritic cells, were evocative, for the presence of the protein p24 in the cytoplasm of these cells. However, the presence of HIV particles between the extensions of these cells in the germinal centers, clearly demonstrated in a number of ultrastructural studies [1, 10–12, 14, 25, 29], suggested that this was possibly a simple staining of these extracellular particles. However, the extent of the staining was highly suggestive of the presence of the protein p24 in the cytoplasm of follicular dendritic cells. So it seems that these cells are infected by HIV and act not only by concentrating viral particles as if they were simple immune complexes in the germinal centers, but also as virus producers, maintaining the infection. Identical observations involving anti-p24 monoclonal antibodies have been noted by other authors [4, 6, 25, 30, 31, 35].

Other cell types containing p24 could be detected, but this was exceptional. These are tingible body macrophages in germinal centers with an intracytoplasmic positivity which is difficult to interpret. Is this proof of viral production or only of phagocytosis of viral particles?

Finally, some small lymphoid cells show clear intracytoplasmic positivity, either in the germinal centers or in the extrafollicular cortical areas. These cells represent the few infected CD4-positive T lymphocytes.

Baroni et al. [4] described the presence of the p24 protein in the cytoplasm of endothelial cells in the lymph node. We were unable to discover such immunolabelled endothelial cells in our cases.

Tenner-Racz et al. [31, 32] observed cells reacting with monoclonal antibodies to HIV-1 gag proteins in the marginal sinuses of lymph nodes. Some of these cells were large, others resembled lymphocytes. They also found strongly labelled large cells between the marginal sinus and the follicles, as well as several positive elongated cells scattered within the parenchyma, particularly around the postcapillary venules which could have represented interdigitating cells [32]. However, they stressed the negativity of the endothelial cells and of the lymphocytes migrating through the vessel wall [32]. Furthermore, they found strong positivity in the germinal centers. In situ hybridization with 32S-RNA probe demonstrated the same topography of positive cells [32].

Conclusion

Immunohistochemical studies of lymphoid tissue provide important data for the diagnosis, prognosis and comprehension of the pathogenesis of HIV infection. The results can be summarized as follows: (1) The most important finding concerning the T lymphocytes is the increased number of CD8-positive cells in the germinal centers of lymph nodes and spleens and in the lymph node sinuses. (2) In addition, a progressive decrease in CD4-positive lymphocytes can be observed in the follicles, in the deep cortex of the lymph node and in the periarteriolar lymphoid sheets of the spleen. This leads to a severe depletion in patients with advanced diseases. (3) A severe B cell hyperplasia with polyclonal plasmacytosis is present in most of the cases. (4) One of the most important lesions was constituted by the progressive fragmentation of the follicular dendritic cells network. This alteration was less marked in the spleen than in the lymph node. (5) The use of monoclonal antibodies against such viral proteins as p24 demonstrated the small number of positive lymphocytes in the lymph node and spleen. Other types of cell seemed to be infected: macrophages and follicular dendritic cells in germinal centers, interdigitating or Langerhans' cells in lymph node sinuses. These findings lead to the hypothesis that the lymph node can be infected through the lymphatics and that the viruses are concentrated and probably produced continuously by the follicular dendritic cells.

References

1 Armstrong, J. A.; Horne, R.: Follicular dendritic cells and virus-like particles in AIDS-related lymphadenopathy. Lancet *ii:* 370–372 (1984).

2 Audouin, J.; Le Tourneau, A.; Marche, C.; Diebold, J.: Etude immunopathologique des adénopathies persistantes liées à l'infection LAV-HTLV III. Ann. Pathol. *6:* 271–274 (1986).

3 Baroni, C. D.; Pezella, F.; Stoppacciario, A.; Mirolo, M.; Pescarmona, E.; Vitolo, D.; Cassano, A. M.; Barsotti, P.; Nicoletti, L.; Ruco, L. P.; Uccini, S.: Systemic lymphadenopathy in intravenous drug abusers. History, immunohistochemistry and electron microscopy: pathogenic correlations. Histopathology *9:* 1275–1293 (1985).

4 Baroni, C. D.; Pezella, F.; Mirolo, M.; Ruco, L. P.; Rossi, G. B.: Immunohistochemical demonstration of p24-HTLV III major core protein in different cell types within lymph nodes from patients with lymphadenopathy syndrome. Histopathology *10:* 5–13 (1986).

5 Biberfeld, P.; Porwit-Ksiazek, A.; Böttiger, B.; Morfelt-Månsson, L.; Biberfeld, G.: Immunohistopathology of lymph nodes in HTLV-III infected homosexuals with persistent adenopathy or AIDS. Cancer Res. *45:* suppl., pp. 4665–4670 (1986).

6 Biberfeld, P.; Chaty, K. J.; Marselle, L. M.; Biberfeld, G.; Gallo, R. C.; Harper, M. E.: HTLV-III expression in infected lymph nodes and relevance to pathogenesis of lymphadenopathy. Am. J. Path. *125:* 436–442 (1986).

7 Biberfeld, P.; Öst, $A.; Porwit, A.; Sandstedt, B.; Pallesen, G.; Böttiger, B.; Morfelt-Månsson, L.; Biberfeld, G.: Histopathology and immunohistology of HTLV-III related lymphadenopathy and AIDS. Acta pathol. microbiol. scand. A *95:* 47–65 (1987).

8 Brask, S.; Hager, H.; Pallesen, G.; Porwit, A.; Biberfeld, P.; Gerstoft, J.: Quantification of CD8-positive lymphocytes in lymph node follicles from HIV-infected male homosexuals and controls. Acta pathol. microbiol. scand. A *95:* 155–157 (1987).

9 Chan, W. C.; Brynes, R. K.; Spira, T. J.; Banks, P. M.; Thurmond, C. C.; Ewing, E. P., Jr.; Chandler, F. W.: Lymphocyte subsets in lymph nodes of homosexual men with generalized unexplained lymphadenopathy. Archs Pathol. Lab. Med. *109:* 133–137 (1985).

10 Diebold, J.; Marche, C.; Aubert, J. P.; Audouin, J.; Le Tourneau, A.; Bouton, C.; Reynes, M.; Wizniak, J.; Capron, F.; Tricottet, V.: Lymph node modification in patients with the acquired immunodeficiency syndrome (AIDS) or with AIDS-related complex. A histological immunohistopathological and ultrastructural study of 45 cases. Pathol. Res. Pract. *180:* 590–611 (1985).

11 Diebold, J.; Le Tourneau, A.; Rousselet, M. C.; Audouin, J.: Histopathologic, immunohistochemic and ultrastructural study of lymph node and spleens in HIV-positive patients at the early phase of the infection. Lymphology *21:* 22–27 (1988).

12 Diebold, J.; Audouin, J.; Le Tourneau, A.; Aubert, J. P.: Lymph node biopsy in the diagnosis of persistent lymphadenopathy syndrome in patients at risk for acquired immunodeficiency syndrome. Progress in surgical pathology, vol. VIII, pp. 55–80 (Field & Wood, Philadelphia 1988).

13 Janossy, G.; Pinching, A. J.; Bofill, M.; Weber, J.; McLaughlin, J. E.; Ornstein, M.; Ivory, K.; Harris, J. R. W.; Favrot, M.; MacDonald-Burns, D. C.: An immunohistological approach to persistent lymphadenopathy and its relevance to AIDS. J. clin. exp. Immunol. *59:* 257–266 (1985).

14 Le Tourneau, A.; Audouin, J.; Diebold, J.; Marche, C.; Tricottet, V.; Reynes, M.: LAV-like viral particles in lymph node germinal centers in patients with the persistent lymphadenopathy syndrome and the acquired immunodeficiency syndrome-related complex: an ultrastructural study of 30 cases. Human Pathol. *17:* 1047–1053 (1986).

15 Marche, C.; Neguesse, Y.; Bouton, C.; Kernbaum, S.; Regnier, B.; Saimot, A. G.; Diebold, J.: Histopathological studies of lymphadenopathy in AIDS: tentative classification. Preliminary report in Epidemic of AIDS and Kaposi's Sarcoma. Symposium, Napoli, vol. 1 (Karger, Basel 1983).

16 Meyer, P. R.; Ormerod, L. D.; Osborn, K. G.; Lowensteiner, L. J.; Hendrickson, R. V.; Modlin, R. L.; Smith, R. E.; Gardner, M. D.; Taylor, C. R.: An immunopathologic evaluation of lymph nodes from monkey and man with acquired immune deficiency syndrome and related conditions. Hematol. Oncol. *3:* 199–210 (1985).

17 Modlin, R. L.; Hofman, F. M.; Meyer, P. R.; Vaccaro, S. A.; Ammann, A. J.; Conant, M. A.; Rea, T. H.; Taylor, C. R.: Altered distribution of B and T lymphocytes in lymph nodes from homosexual men with Kaposi's sarcoma. Lancet *i:* 758–771 (1983).

18 Muller, H.; Falk, S.; Stutte, H. J.: Accessory cells as primary target of human immunodeficiency virus HIV infection. J. clin. Path. *39:* 1161–1170 (1986).

19 Muller, H.; Falk, S.; Stutte, H. J.: Immunohistochemical investigations of lymph nodes in LAS and AIDS patients. Cancer Detect. Prevent. *1:* suppl., pp. 577–582 (1987).

20 Öst, Å.; Baroni, C. D.; Biberfeld, P.; Diebold, J.; Moragas, A.; Noel, H.; Pallesen, G.; Racz, P.; Schipper, M.; Tenner-Racz, K.; Van den Tweel, J. G.: Lymphadenopathy in HIV infection: histological classification and staging (in press).

21 Pallesen, G.; Gerstoft, J.; Mathiesen, L. R.; Dick-Meiss, E.; Platz, P.; Kroon, S.; Linhardt, B.; Brask, S.; Pedersen, N. S.: Histological and immunohistological lymph

node changes in homosexual men are specific for HTLV-III/LAV infection and correlate with clinical and immunological findings. Pathol. Res. Pract. *180:* 304–312 (1985).

22 Pallesen, G.; Gerstoft, J.; Mathiesen, L. R.: Stages in LAV/HTLV-III lymphadenitis. I. Histological and immunohistological classification. Scand. J. Immunol. *25:* 83–88 (1987).

23 Pekovic, D. D.; Gornistsky, M.; Ajdukovic, D.; Dupuy, J. M.; Chausseau, J. P.; Michaud, J.; Lapointe, N.; Gilmore, N.; Tsoukas, C.; Zwadlo, G.; Popovic, M.: Pathogenicity of HIV in lymphatic organs of patients with AIDS. J. Pathol. *152:* 31–35 (1987).

24 Pileri, S.; Rivano, M. T.; Raise, E.; Gualandi, G.; Gobbi, M.; Martuzzi, M.; Gritti, F. M.; Gerdes, J.; Stein, H.: The value of lymph node biopsy in patients with the acquired immunodeficiency syndrome (AIDS) and the AIDS-related complex: a morphological and immunohistochemical study of 90 cases. Histopathology *10:* 1107–1129 (1986).

25 Racz, P.; Tenner-Racz, K.; Kahl, C.; Feller, A. C.; Kern, P.; Dietrich, M.: The spectrum of morphologic changes of lymph nodes from patients with AIDS or AIDS-related complexes. Prog. Allergy, vol. 37, pp. 81–181 (Karger, Basel 1986).

26 Raphael, M.; Pouletty, P.; Cavaille-Coll, M.; Rozenbaum, W.; Homon, D. A.; Nonnenmacher, L.; Delcourt, A.; Gluckman, J. C.; Debre, P.: Lymphadenopathy in patients at risk for acquired immunodeficiency syndrome. Histopathology and histochemistry. Archs Pathol. Lab. Med. *109:* 128–132 (1985).

27 Rousselet, M. C.; Audouin, J.; Le Tourneau, A.; Bouchard, I.; Espinoza, P.; Kazatchkine, M.; Diebold, J.: Idiopathic thrombocytopenic purpura in patients at risk for acquired immunodeficiency syndrome: histopathology, immunohistochemistry and ultrastructural study on six spleens. Archs Pathol. Lab. Med. *112:* 1242–1250 (1988).

28 Said, J. W.; Shintaku, I. P.; Teltelbaum, A.; Chien, K.; Sassoon, A.: Distribution of T cell phenotypic subsets and surface immunoglobulin-bearing lymphocytes in lymph nodes from male homosexuals with persistent generalized adenopathy: an immunohistochemical and ultrastructural study. Human Pathol. *15:* 785–790 (1984).

29 Tenner-Racz, K.; Racz, P.; Dietrich, M.; Kern, P.: Altered follicular dendritic cells and virus-like particles in AIDS and in AIDS-related lymphadenopathy. Lancet *i:* 105–106 (1985).

30 Tenner-Racz, K.; Racz, P.; Bufill, M.; Schultz-Meyer, A.; Dietrich, M.; Kern, P.; Weber, J.; Pinching, A. J.; Veronese-Dimarzo, F.; Popovic, M.; Klatzmann, D.; Gluckman, J. C.; Janossy, G.: HTLV-III/LAV viral antigens in lymph nodes of homosexual men with persistent generalized lymphadenopathy and AIDS. Am. J. Path. *123:* 9–15 (1986).

31 Tenner-Racz, K.; Racz, P.; Dietrich, M.; Kern, P.; Janossy, G.; Veronese-Dimarzo, F.; Klatzmann, D.; Gluckman, J. C.; Popovic, M.: Monoclonal antibodies to human immunodeficiency virus: their relation to patterns of lymph node changes in persistent generalized lymphadenopathy and AIDS. AIDS *1:* 95–104 (1987).

32 Tenner-Racz, K.; Racz, P.; Schmidt, H.; Dietrich, M.; Kern, P.; Louie, A.; Gartner, S.; Popovic, M.: Immunohistochemical, electron microscopic and in situ hybridization evidence for the involvement of lymphatics in the spread of HIV-1. AIDS *2:* 299–309 (1988).

33 Toccanier, M. F.; Kapanci, Y.: Lymphadenopathy in drug addicts. A study of the distribution of T-lymphocyte subsets in the lymph nodes. Virchows Arch. A Path. Anat. Histopathol. *406:* 149–164 (1985).

34 Turner, R. R.; Meyer, P. R.; Taylor, C. R.: Immunohistology of persistent generalized

lymphadenopathy. Evidence of progressive lymph node abnormalities in some patients. Am. J. clin. Pathol. *28:* 10–19 (1987).

35 Ward, J.M.; O'Leary, T.J.; Baskin, G.B.; Benveniste, R.; Harris, C.A.; Nara, P.; Rhodes, R.H.: Immunohistochemical localization of human and simian immunodeficiency viral antigens in fixed tissue sections. Am. J. Path. *127:* 199–205 (1987).

36 Wood, G.S.; Garcia, C.F.; Dorfman, R.F.; Warnke, R.A.: The immunohistology of follicle lysis in lymph node biopsies from homosexual men. Blood *66:* 1092–1097 (1985).

J. Audouin, MD, Service Central d'Anatomie et de Cytologie Pathologiques, Hôtel-Dieu, 1, Place du Parvis Nôtre-Dame, F–75181 Paris Cédex 4 (France)

Racz P, Haase AT, Gluckman JC (eds): Modern Pathology of AIDS and Other Retroviral Infections. Basel, Karger, 1990, pp 201–221

Double Immunofluorescence Analysis of Lymphoid Tissues during HIV-1 Infection[1]

George Janossy[a], *Margarita Bofill*[a], *Paul Racz*[b]

[a]Department of Immunology, Royal Free Hospital School of Medicine, London, UK; [b]Department of Pathology, Bernhard Nocht Institute for Tropical Medicine, Hamburg, FRG

There are two basic unsolved problems in understanding the pathogenesis of HIV-1 infections leading to the development of acquired immune deficiency syndrome (AIDS). *First,* in the tissues of patients with asymptomatic HIV-1 infections only a few infected lymphocytes and macrophages can be identified. The estimates in the lymph nodes vary depending upon the methods used, but both the analysis of p24 core HIV-1 antigen expression and the in situ hybridization methods indicate that <0.1–0.01% of cells appear to be infected [1–6]. The recent introduction of polymerase chain reaction (PCR) has increased the sensitivity of detecting HIV-1 involvement and can also be interpreted semiquantitatively in terms of the numbers of infected cells present when applied in limiting dilution analysis [7, 8]. As yet, only results from the peripheral blood and bronchoalveolar lavages have been reported but these again indicate that despite such increased sensitivity the cells with identifiable HIV-1 gag and env gene sequences represent only 0.1–1% of the recirculating cells [7, 8]. The functional alterations seen in short-term cultures of cells taken from HIV-1-infected asymptomatic individuals are more profound than it would be expected from the involvement of relatively minor proportions of cells [9, 10], indicating that, in addition to the direct destruction of $CD4^+$ lymphocytes by HIV-1, indirect immunopathological mechanisms may also contribute to the relentless progression of this disease. A number of such secondary events have been envisaged [11–14] (see below), and immuno-

[1]This collaborative work was supported by the Wellcome Trust (UK) and by the Körber Foundation, Hamburg, and BMFT (Bundesministerium für Forschung und Technologie), Bonn, FRG.

histological studies should give definite clues in confirming or refuting these working hypotheses.

Second, the progression to AIDS is a long process with features of apparent immunostimulation [15–17] and anergy of unusual severity [cf. 18]. The knowledge about the exact cellular interactions which lead from a well-preserved, although apparently activated, lymph node architecture towards a devastating involution of all lymphoid tissues including central and peripheral lymphoid organs such as the thymus [19], bone marrow [20], lymph nodes [15–17], spleen [21] and gut [22] should not only illustrate the prominent sequence of events for understanding retroviral diseases in general, but also give a definite ray of hope for preventing such deterioration of immunity in asymptomatic HIV-1-positive people.

The most logical and informative avenue for approaching the two main basic questions outlined above is through the investigation of the immunohistology of HIV-1 infection. This is for three rather obvious reasons:

(1) Although not as easily accessible as blood cells, it is in the lymphoid tissues and the brain [23] where the brunt of retroviral attacks takes place. Blood cells can only be regarded as convenient but haphazard and incomplete samples of the tissue bound cellular populations. On the one hand, at least two crucial cell types are absent from the blood: the well-developed forms of the macrophage lineage [24] and those of follicular dendritic cells [25]. On the other hand, cell types including the whole family of 'natural killer' cells exert their actions primarily in the blood and are either absent or altered, both phenotypically and functionally, in the tissues.

(2) There is a virtually complete agreement about the major immunohistological changes seen during the progress of HIV-1 infection into AIDS [15–17, 26, reviewed in 27]. In these studies the informative content of observations obtained by histology has been enhanced by the use of monoclonal antibodies (MAbs) to T lymphocyte subsets, B lymphocytes and accessory cells such as macrophages and follicular dendritic cells (table 1) and these changes will serve as the basis for further investigations (see below). Nevertheless, an important practical issue is that during these studies it has been sufficient to apply only single colour immunohistology, e.g. immunoperoxidase, without a need for the combined use of multiple immunological markers in a single preparation.

(3) There has been an impressive recent development, fuelled by the combined effects of monoclonal antibody and cytokine revolutions, in the functional analysis of immunoregulatory circuits [cf. 28, 29]. These advancing branches of science use different technologies but converge in the multi-

Table 1. The basic set MAbs for single colour immunohistology[1]

Antibody	Specificity	Comment
CD3	pan-T	functional T cell antigen associated with T cell receptor
CD4	T cells of helper type	reacts with T cells which see antigens in context of class II
CD8	T cells of suppr/ctx type	reacts with T cells which see antigens in context of class I
CD20	pan-peripheral B cell	reacts with B cells and FD cells (v. weakly)
Anti-IgM	most B cells	reacts with both SmIg and complexes in the germinal centres
Anti-IgD	recirculating B cells	clear membrane staining of B cells in lymphocyte corona
KiM4, RFD3 or anti-C3b	FD cells	react with most FD cells; anti-C3b shows the most extensive staining
CD68	pan-macrophage	reacts with monocytes and tissue macrophages (e.g. reagents KiM7 and EBM11)

[1]Reviewed in Racz et al. [27] and Diebold et al. [this volume].

parameter analysis of various cell populations. The volume and diversity of new information appear to be confusing and the various cytokines are pleiotropic, but this confusion is most probably merely due to the fact that the microenvironmental aspects of local cytokine production and lymphocyte/accessory cell function have not yet been rigorously studied. For this reason it is difficult, at the moment, to discriminate between important cellular interactions and in vitro artefacts – which might be reproducible but are of no (patho)physiological consequence. The same conclusion applies to the pathogenesis of HIV-1 infections: the altered microenvironments have not yet been dissected with modern technology. The crucial issue here is that, in contrast to conventional studies, these investigations require the routine use of multiple markers applied in multicolour combinations.

The aim of this short review is therefore threefold. First, we refer to the accepted view of histological changes within the HIV-1-infected lymphoid tissues in order to define the questions of disease progression in immunohistologic terms. Second, we describe the methods for double and triple labeling using reagents frequently applied in such analysis. Finally, three recent observations about the biosynthetic activities of lymphoid subsets are highlighted for depicting the changing balance in HIV-1-infected lymphoid tissues which may contribute to secondary pathogenetic mechanisms leading to AIDS.

Histological Changes in HIV-Infected Lymphoid Tissues

The successive histological changes of lymphadenitis in HIV-1 infection are frequently associated with (a) follicular *hyperplasia;* (b) a *mixed pattern* revealing signs of hyperplasia and involution together in the same lymph nodes with increasing evidence of follicular lysis leading to the formation of lacunar spaces [15–17, 27] and the appearance of hyaline deposits [30], and (c) *lymphocyte depletion* completing the process of involution [15–17, 27]. Transition from one to another type usually takes a long time: such a change may not be observed during the observation period of >4 years [30]. For diagnostic reasons, patients with persistent generalized lymphadenopathy (PGL) have been studied most frequently; they represent 50–55% of HIV-1^{+} individuals. Nevertheless, there are two other variants of presentation. In one of the other major group the asymptomatic patients show less prominent follicular expansion but the few biopsies studied revealed the other main histological features (fig. 1). Finally, in about 10% of lymph node samples the lymphadenitis is associated with additional signs of hypervascularization which is facilitated by HIV-1-infected CD4^{+} lymphocytes secreting an endothelial growth factor as a predisposing cytokine for the development of Kaposi sarcoma [31]. The reasons for such variations in the presentation of HIV-1 lymphadenitis are unknown.

The main features of lymph node reaction in the typical cases of PGL, analysed by a reagent panel similar to that shown in table 1, are shown in figure 1. In essence, the enlargement of *follicles* can be rather extreme, resulting in bulging germinal centres of bizarre serpentine and hourglass shape. There are many dividing centroblasts and abundant cellular debris with increased numbers of phagocytic ‘tingible body’ macrophages, particularly in the earlier stages. The area of centrocytes, some showing plasmacytoid features, is also expanded. At the same time the lymphocyte corona is frequently thin or even absent in the plane of the section. Inside these germinal centres p24 core antigen is regularly deposited in large amounts [32]: such p24 is visibly complexed to immunoglobulin on the surface of follicular dendritic (FD) cells and is likely to represent viral antigens released from destroyed particles (fig. 2c). Whether or not free p24 antigen is synthesized or taken up

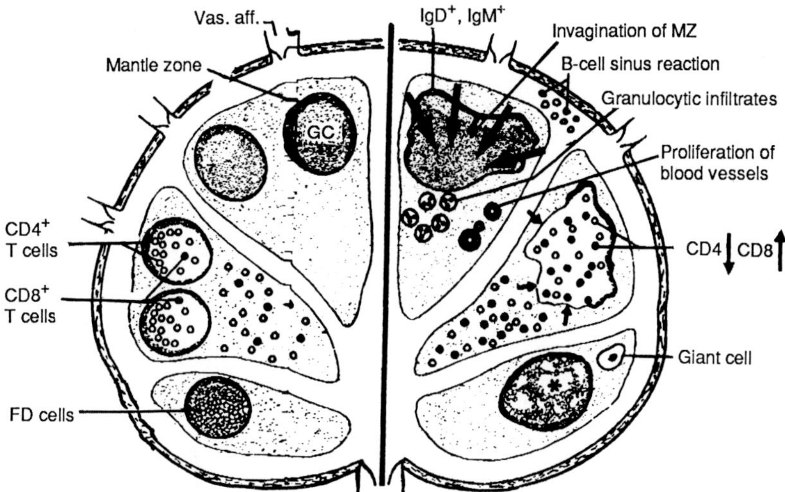

Fig. 1. The main features of lymphadenitis with follicular hyperplasia. The diagram depicts the morphology of control lymph node (left) in comparison with the major changes in HIV-1^{+}-infected nodes (right): the enlargement of follicles with germinal centres (GC), the deposition of p24 core antigen (large arrows) and the eroded or attenuated lymphocyte corona (small arrows). In such GCs increased numbers of CD8^{+} and decreased numbers of CD4^{+} cells are detected. At later stages the FD cell networks inside the GCs reveal a moth-eaten appearance (asterisk). In the extrafollicular parenchyma a lower than normal CD4/CD8 ratio, small clusters of granulocytes, proliferation of blood vessels and often multinucleated giant cells are seen. The mantle zone (MZ) is also referred to as lymphocyte corona. Modified from Racz et al. [27], with permission of the publishers.

inside the FD cells remains to be proven but intact viral particles can be unequivocally demonstrated by electron microscopy, particularly in the vicinity of infiltrating CD8^{+} lymphocytes. These viral elements may be extremely important in reinfecting passing CD4^{+} T cells but are likely to represent only a fraction of the total p24 antigen load which is detectable by anti-p24 antibodies inside the germinal centres. The normal constituents of follicles are CD4^{+} T lymphocytes: these decrease in number with the progression of HIV-1 infection although may not vanish totally even when the blood CD4 counts are low [33]. There is, on the other hand, a powerful influx of CD8^{+} cells into the HIV-1^{+} germinal centres where they are present in >12 times higher concentrations than in the corresponding areas of normal lymph nodes [33–35]. The *extrafollicular lymphoid parenchyma* shows the strain of the gradual depletion of CD4^{+} T cells with variable but mostly inverted CD4/CD8 ratios. There are oedematous changes, many congested vessels with swelling of the high endothelial venules. The accumulation of plasma cells is quite remarkable in many lymph nodes despite $CD4^+$ depletion: these plasma cells are such regular constituents of the severely involuted nodes that one wonders where their progenitor B cells are and that how long these plasma cells may live. The *sinusoids* show an enhanced cellularity, frequently with many polymorphs and sometimes with accompanying activated B lymphoblasts which cause sinus histiocytosis at this site. In these areas, HIV-1^+ macrophages can also be found indicating that the HIV-1 infection spreads through the lymphatic system [36].

Technical Aspects of Marker Combinations

The use of multiple immunohistological markers in AIDS research is necessary in the following areas: (a) in investigating the microenvironmental relationship of various lymphoid/accessory cell populations; (b) the clarification of the phenotypes of rare cells which express various viral proteins and/or viral RNA or DNA; (c) the assessment of functional features which require two or more markers for appropriate analysis (table 2). During the last 6 years it has become clear that the lineage-specific MAbs need to be applied together

---

Fig. 2. Two colour IF using heteroantisera and -antibodies prepared in different species. In these sections areas of germinal centres are depicted using labeling of human IgM with G-anti-Hu-IgM-peroxidase (*a*, 'conventional' control), G-anti-Hu-IgM-FITC (*b*, green) or G-anti-Hu-IgM-TRITC (*c–f*, red). The samples shown are hyperplastic lymph nodes with PGL from HIV-1^+ patients in *a–d*, an involuting HIV-1^+ lymph node in *e* and a control sample of normal tonsil from a child in *f*. The second labels and the interpretation of observations are as follows. In *b* CD8 plus G-anti-M-Ig-TRITC is used to demonstrate the entry of $CD8^+$ cells into the GC below the lymphocyte corona. At the top the paracortical zone is shown with abundance of $CD8^+$ cells. In *c* the green indirect label is M-anti-p24 MAb plus G-anti-M-Ig-FITC. In the middle of the GC the IgM^+ antigen-antibody complexes (red) contain P24 antigen (green) and the double exposure yields a yellow colour. Uncomplexed p24, if present, is in relatively small quantities. The rest of the plate *(d–f)* is double labled with CD23 plus G-anti-M-Ig-FITC. This functional antigen, which is a growth factor for B cells, is present in very large amounts on the FD cells throughout the GC in PGL *(d)* and even the remnants of FD cells during involution are strongly $CD23^+$ *(e)*. As a marked contrast, in the GC of normal tonsil the $CD23^+$ FD cells are restricted to the apical light zone (the middle of field in *f*) while the FD cells in the lower part (in the 'dark' zone) are $CD23^-$. Magnification is $100\times$ in *a* and $200\times$ in *b–f*.

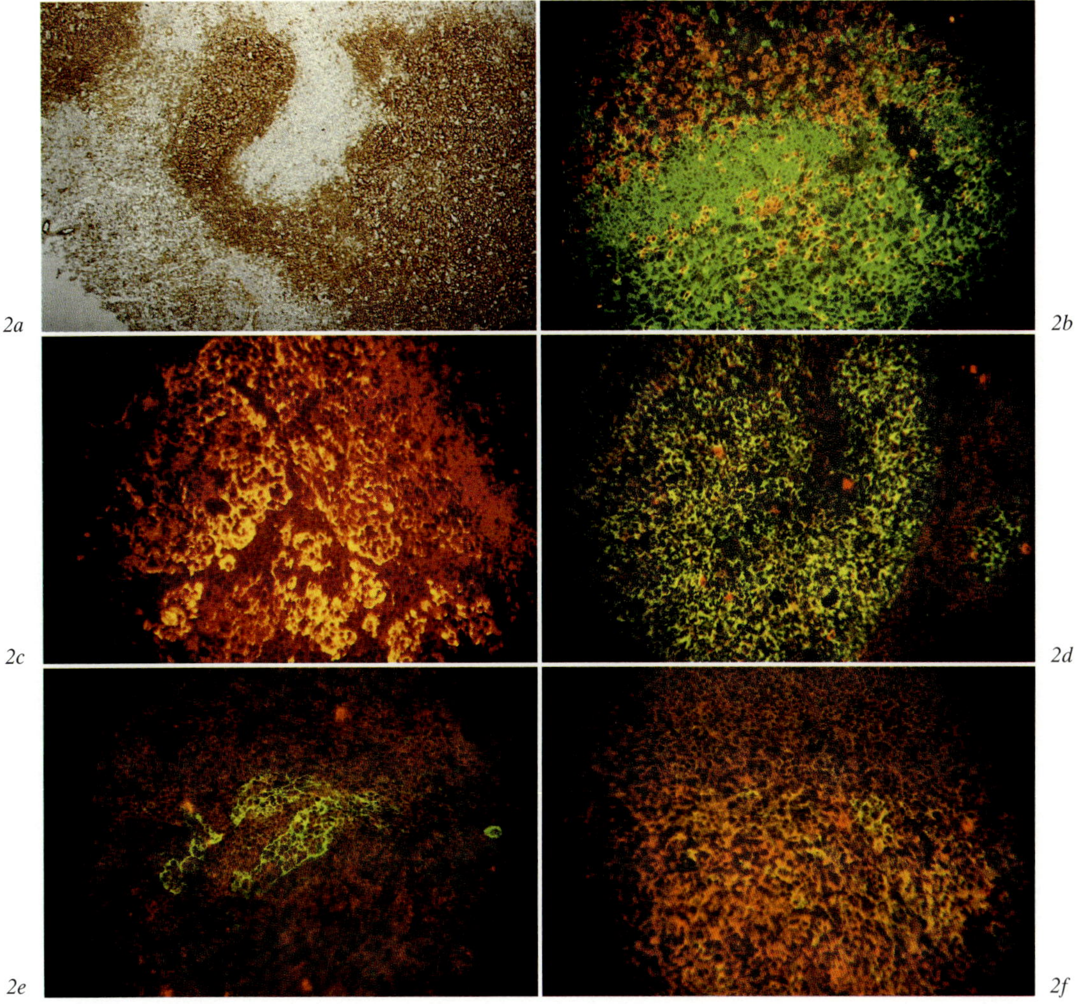
2a
2b
2c
2d
2e
2f

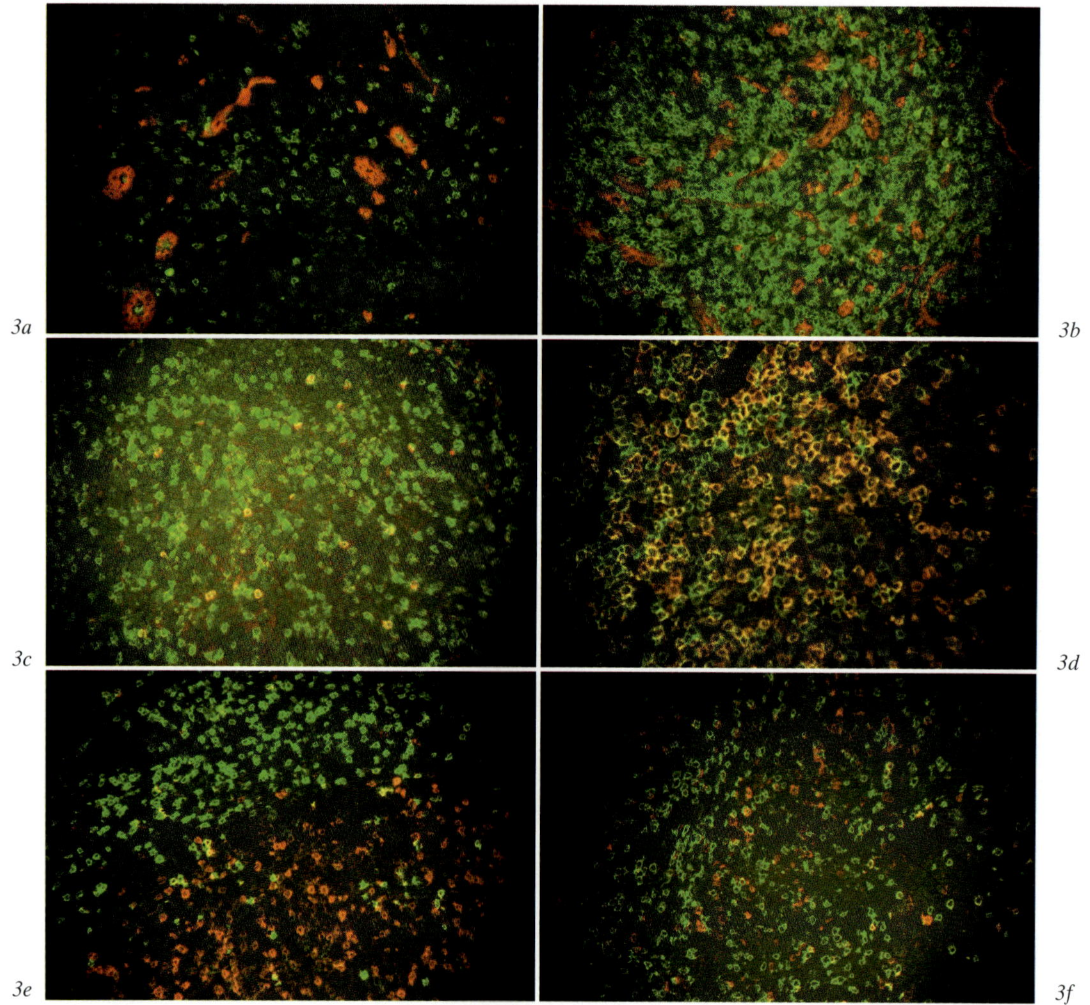
3a
3b
3c
3d
3e
3f

with markers which on their own are not lineage specific but yield invaluable data about additional functional aspects [37]. Two examples of this general concept are mentioned here. Firstly, there are new efforts to make MAbs to functional molecules such as interferons and interleukins and use them in assessing the synthetic activities of the various cell types in situ [38, 39]. More and more MAbs are becoming available to integrins [40] and endothelins [41] with clearer functional connotations. Previously standardized function-associated reagents are reinterpreted in more precise molecular terms [42, 43]. Secondly, the consecutive changes of development such as the transition from immunologically naive or unprimed 'virgin' cells to experienced primed 'memory' cells have recently been described within the various T cell lineages (see below). The relevant points here are that the 'markers' for naive T lymphocytes, including the CD45RA molecules first detected by the HB11 and 2H4 MAbs [44, 45], are not T cell specific but present on B lymphocytes and haemopoietic precursor cells. Similarly, the markers for 'memory' T cells including the CD45RO molecule [46] are present not only on T cells but also on subsets of macrophages. The functional characteristics of the 'virgin' and 'memory' T cells are divergent (see below). Hence the need for staining combinations in detecting CD45R antigens on $CD4^+$ and $CD8^+$ T lymphoid cells (fig. 3c, d).

Fig. 3. Comparative phenotypic analysis of $CD8^+$ cells in the normal tonsil *(a, c, e)* and in lymphadenitis during HIV-1 infections *(b, d, f)*. In this figure two colour immunofluorescence with MAbs of different classes is demonstrated. In *a–d* the CD8 used is of IgM class and other MAb is of IgG class. In *e* and *f* the CD8 is an IgG in combination with a reagent of IgM class. Throughout CD8 Abs are labeled with FITC-conjugated second layers (green) while the partner MAbs are labeled with TRITC-conjugated second layers (red). The interpretation of observations are as follows: in *a* and *b* CD8 labeling is combined with a MAb to factor VIII to illustrate blood vessels in the paracortical area. The accumulation of $CD8^+$ and endothelial cells in lymphadenitis is clearly shown. In *c* and *d* CD8 labeling is combined with UCHL1 detecting memory cells (CD45RO in ref. 28). In the normal tonsil a large proportion of $CD8^+$ cells (green in *c*) are $UCHL1^-$ and only a few $CD8^+$, $UCHL1^+$ doubles are seen *(c)* while in HIV-1^+ lymphadenitis the majority of $CD8^+$ cells are doubly labeled for UCHL1. This is shown in *d* with higher magnification. Finally, in *e* and *f* CD8 labeling is combined with Leu7 and the paracortical areas are depicted on the top left and the germinal centres are depicted in the middle and bottom right. In the tonsil *(e)* $CD8^+$ cells remain largely restricted to the paracortex and the $Leu7^+$ cells (which are mostly $CD4^+$; not shown) are in the GC. In HIV-1^+ lymphadenitis the $CD8^+$ cells invade the GC and mostly $Leu7^-$ and only few $Leu7^+$, $CD8^-$ cells are left. The possibility of creating an imbalance of cytokine secretion is discussed in the text and in figure 6.

Table 2. Markers utilized to characterize functionally different populations within the main categories[1]

Category	Additional aim	Marker	Frequently used Ab	Reference
Pan-T or T subset (CD4⁺ or CD8⁺)	a) TCR expression[2]	TCRαβ	βF1 and WT31	48
		TCRγδ	TCRδ1	49
	b) virgin and memory T cells	CD45RA (virgin)	2H4, HB11 or Sn130	45, fig. 5
		CD45R0 (memory)	UCHL1	46, fig. 3c, d, 5
	c) lymphokine synthesis	e.g. IFNγ, IL-1	IGNB-14	38, 39
NK-like	a) genuine NK and ADCC effectors	low aff.FcR	Leu11a	50
	b) other subsets of unknown significance	endothelin family	Leu7	51, fig. 3e, f
B cells	a) Ig (sub)class	Abs to IgM, IgD, IgA and IgG		fig. 2b–f
	b) subsets inside the follicles	integrins	Leu8	52
		growth factors	CD23	53, fig. 2d–f
Accessory cells	a) FD cells	growth factors	CD23	53, fig. 2d–f
	b) subsets of macrophages (recently generated)	CD14	UCHM1	24
			RFD7	24
	c) interdigitating and Langerhans' cells	HLA-DQ related:	Tü22 & RFD1	24
		CD1	Na1/34	54
Proliferation	cells in G1-S-G2-M phases	nuclear Ag	Ki67	55

[1] Main cell types are shown in table 1.
[2] TCR = T cell receptor.

The progressive changes in this area of technology are driven by the recent advances of flow cytometry for the following reasons. The great improvement in the sensitivity of fluorescence-activated flow cytometry and the availability of five parameter analysis which investigates three fluorescence channels with a single laser elicitation has not only introduced the combination analysis into the routine laboratory but also continues to provide reagents of very high quality. These are available as directly conjugated reagents in three different forms: conjugated to fluorescein-isothiocyanate (FITC), phycoerythrin (PE) and duochrome (DC) [reviewed in 47]. The important issue for tissue analysis, which can hardly be overemphasized, is that such a five parameter analysis using three fluorescence markers and two scatter channels in combination [47] is eminently *economical* and precise when minute tissue samples such as lymphoid populations teased out from small biopsies are being investigated. Thus these studies ideally complement immunohistology performed in the same or in a parallel sample.

The possible permutations for scientifically interesting combinations are manifold and some of the informative examples are demonstrated in figures 2 and 3. There are four relevant points for discussion in table 3:

(1) Obviously, combinations can be most conveniently performed by directly labelled reagents: these assays are both rapid and easy to perform because the chances for irrelevant cross-reactions between the reagents are minimal. Nevertheless, in immunohistology these direct assays can only be satisfactorily performed with the strongly expressed antigens, e.g. immunoglobulins (fig. 2b–f), CD8 and a few other antigens such as CD45RA (e.g. 2H4-PE). Since these MAbs are purchased in conjugated form and have to be used fairly lavishly, the direct assays can be expensive.

(2) One of the most powerful and economical methods for staining combinations utilizes MAbs of different (sub)classes with conjugated (sub)class-specific second layers. The CD groups of antibodies standardized at the Leucocyte Typing Workshops almost invariably contain reagents of varying (sub)classes, and affinity purified specific goat antibodies to murine IgM, IgG, IgG1 and IgG2 are commercially available, e.g. from Southern Biotechnological Associates. These second layer reagents are conjugated to different fluorochromes (table 3) and can be conveniently used in various modifications with a wide variety of first layer MAbs. Three combinations using two different CD8 antibodies and various second layers are depicted in figure 3. The particular advantages of these methods are their versatility and economy. Fortunately the second layers are of such high quality that the signal amplification is impressive and the cross-reactions are avoided.

Table 3. Practical double and triple immunofluorescence (IF) combinations[1]

Feature	Example	First layer	Second layer	Reference
Direct labeling	anti-Hu-Ig	G-anti-Hu-IgM-FITC	none	fig. 2b
		G-anti-Hu-IgM-TRITC	none	fig. 2c–f
	Mab to CD45RA	2H4-PE	none	Coulter catalogue
Indirect labeling with MAbs of different (sub) class	CD8	of IgM class	G-anti-M-IgM-FITC	fig. 3a–d
	CD8	of IgG class	G-anti-M-IgM-FITC	fig. 3e, f
	CD45R0	of IgG class	G-anti-M-IgG-TRITC	fig. 3c, d
	Leu7	of IgM class	G-anti-M-IgM-TRITC	fig. 3e, f
Indirect labeling with MAb biotin-avidin	MAbs conjugated to biotin		avidin or streptavidin-FITC, -TRITC, -PE or -duochrome (DC)[2]	various companies including Becton-Dickinson

[1]Triple labelling is convenient using ABs conjugated to FITC (green), PE (orange) and Duochrome (DC: red) [47]. On the microscope the selective barrier filter has to be removed from the FITC filter set and the green, orange and red colours are readily distinguishable. When laser flow cytometry is applied with a single laser (e.g. on the FACScan) the three channels are evaluable by the separate filter setting. During the microscopical analysis antifading chemicals are necessary (see text).
[2]Streptavidin-PE is from Amersham plc and Streptavidin-DC is from Becton-Dickinson.

(3) Occasionally both MAbs are of the same class and the antigens are weakly expressed. In this situation the staining is sequential according to the following steps. The first MAb is stained with G-anti-M-Ig-FITC and the sections are incubated with blocking mouse serum (1:10 dilution). The second MAb is used in a biotinylated form and labelled with fluorochrome-conjugated streptavidin.

(4) Second layer reagents conjugated with tetramethyl rhodamine-isothiocyanate (TRITC) are not useful for flow cytometry and may be in shorter supply. The relevant comments about the alternative PE- and DC-conjugated reagents are twofold. First, in order to optimally visualize the orange colour of PE the selective barrier filter needs to be removed from the dark green FITC filter set. With this nonselective FITC setting the triple labelling by FITC (green), PE (orange) and DC (red) is clearly visible together, while the TRITC filter set allows the visualization of PE (weakly) and DC (strongly). Second, these new fluorochromes fade so rapidly on the epifluorescence microscope, particularly on TRITC illumination, that the staining can be properly studied only in the presence of antifading compound which is available from Amersham plc as part of the Streptavidin-PE kit.

Clues about the Pathogenesis of Immunoregulatory Disorders in HIV-1 Infections

Some of the clues about the immune disorders in HIV-1 infections can be deciphered from the immunohistological changes (fig. 2, 3). Three observations are discussed sequentially.

The first relevant point is the accumulation of large amounts of p24 gag in the form of immune complexes within the germinal centres (fig. 2c) without a concomitant deposition of envelop proteins such as gp120 [17, 27, 56]. The simplest explanation for this finding is that following viral replication only the p24 is complexed efficiently by anti-p24 antibodies in a manner that leads to their accumulation inside the germinal centres, while gp120 may bind to CD4 molecules and remain dispersed. A corollary of this possibility is that the binding of gp120 to $CD4^+$ cells may interfere with these cells' function. It is known that gp120 and class II antigens are 'seen' by the same second domain of the CD4 molecule [57] and gp120 may indeed alter the $CD4^+$ cells' response to antigens presented in the context of self-class II.

The second interesting issue is the recent advance in understanding the heterogeneity of peripheral T lymphocytes in man. In 1971, Raff and Cantor

[58] formulated the concept about the existence of two distinct populations of T helper cells in the mouse, corresponding to virgin (T1) and memory (T2 cells). In humans the corresponding populations of cells have also been gradually characterized. On the one hand, the seminal studies by Tedder et al. [44] and Morimoto et al. [45] have described two similar MAbs, HB11 and 2H4, which both delineate a functionally distinct 'suppressor-inducer' subset within the T-helper cell populations. Both of these MAbs react with the restricted p200 and p220 polypeptides of the leucocyte common antigen (LCA; CD45) referred to as CD45RA [28]. On the other hand, it has been observed that the complementary populations of $CD4^+$ cells are more accomplished helper cells [42] and exhibit the restricted moieties of LCA of the lower p180 molecular weight. This is recognized by the MAb UCHL1 [46] and referred to as CD45RO [59]. Soon it has been demonstrated that the phenotype of activated $CD45RA^+$ T lymphocytes changes during the consecutive mitoses in culture: these cells lose CD45RA (2H4) positivity while gaining CD45RO (UCHL1) positivity. This apparently irrevocably change has been mapped into the virgin → memory cell switch. A further important point has also been documented: both CD4 and CD8 populations switch their CD45 antigens during this process [60, 61]. Interestingly, following 'memory' T cell development the primed cells exhibit higher levels of adhesion molecules such as LFA-1, LFA-3 and VLA ('very late antigen' [43]) defined by the 4B4 MAb [42], thus increasing these cells' capabilities to form cellular interactions [61]. The altered features of memory cells are also represented in their recirculation patterns. $UCHL1^+$ cells selectively migrate to special sites such as the germinal centres of lymph nodes and the lamina propria in the gut [62] and also seek out areas of immune activation, e.g. inflammed synovium during rheumatoid arthritis and granulomas [reviewed in 63].

These observations are relevant for the development of AIDS for the following reasons. In the HIV-1-infected lymph nodes the most dramatic changes include the accumulation of $CD8^+$ cells in the paracortical area (fig. 1, 3b), and this enrichment is due to the relentless increase in the number of $UCHL1^+$ (memory) $CD8^+$ population (fig. 3d). Similarly, in the germinal centres the newly arriving $CD8^+$ 'invaders' (fig. 2b, 3f) are also >95% $UCHL1^+$ [64]. This is a substantial change, both in absolute number and in phenotype, because in the normal lymph nodes the majority of $CD8^+$ cells carry the markers of 'virginity' ($CD45RA^+$, $UCHL1^-$). When analysed repeatedly in samples taken from the same donor at various times it also becomes clear that these phenotypic shifts within the $CD8^+$ populations occur relatively early during the asymptomatic phase of HIV-1 infection, i.e. many years

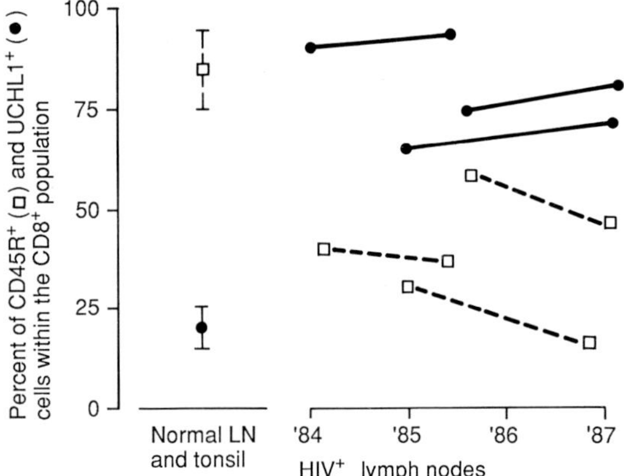

Fig. 4. A longitudinal study of developing UCHL1⁺, CD8⁺ cells in HIV-1-infected lymph nodes. The biopsies were taken from the same individuals as shown. ● = UCHL1⁺ cells; □ = CD45RA⁺ cells within the CD8⁺ populations determined by double IF as shown in figure 3c, d.

before an AIDS-related complex develops (fig. 4) [64]. The possible functional significance of these findings is that when the various T cell subsets are investigated for cytotoxic activity in vitro against target cells sensitive to CD8⁺ T cell mediated killing, the CD8⁺, UCHL1⁺ cells isolated from normal donors show exceptionally high cytotoxicity (fig. 5). In order to achieve such a high killing, the T cells need to be activated (with CD3 MAbs or PHA, etc.) [59]. It remains to be investigated whether or not the CD8⁺, UCHL1⁺ cells isolated from HIV-1-infected lymph nodes are also capable of exhibiting a similar cytotoxic potential. A related question is whether this high level of cytotoxicity, also demonstrated in *HIV-1-specific* cytotoxic assays by cells taken from the blood and the bronchoalveolar lavages of HIV-1-infected individuals [65, 66], protects the individuals from the spread of HIV-1 infection or causes autoaggressive damage. Nevertheless, from an immunohistological point of view, the most credible working hypothesis about the potential role of autoaggressive cells is the facilitation of AIDS by an overactivated and abundant UCHL1⁺, CD8⁺ population which help destroy CD4⁺ cells, perhaps after these autologous targets passively bind and/or 'process' gp120 (see above). Occasional investigations in the literature support the possibility of cytotoxic activity against noninfected CD4⁺ cells [67] but further confirmatory studies, particularly following the preincubation of uninfected histocompatible CD4⁺ cells with gp120, are still warranted.

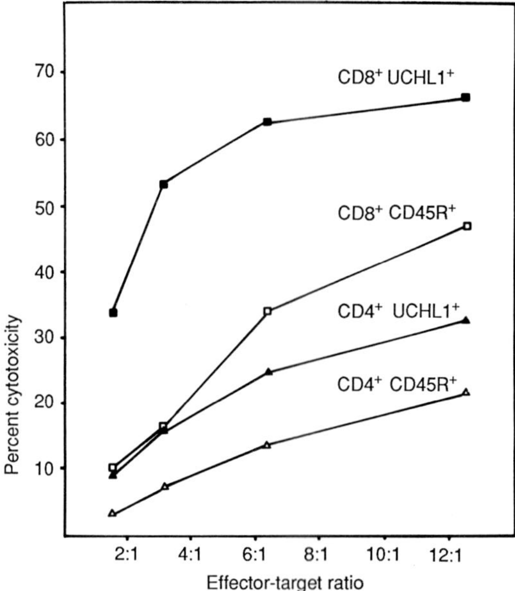

Fig. 5. The cytotoxic capabilities of the 'virgin' and 'memory' T cell populations against a T-cell-sensitive target cell (P815 murine mastocytoma). Following the reactivation of the lymphocytes with nonspecific mitogens the most efficient cytotoxic cells are the $CD8^+$, $UCHL1^+$ (memory) cells while the $CD8^+$, $UCHL1^-$ (virgin) cells show weaker killing effects. The $CD4^+$ cells are even less efficient [59].

The final relevant issue is the evidence recently obtained about the overexpression of CD23 antigens in the HIV-1-infected lymph nodes [64]. Here we have to digress first because CD23 antigen is an interesting molecule which has been discovered more than once as an autochthonous B cell growth factor [68, 69], IgE receptor of low affinity [70] and a B-blast-associated membrane antigen [71]. The molecule is expressed on activated B cells and monocytes; the display on recirculating IgM^+, IgD^+ primary B cells is markedly up-regulated following activation by stimuli such as EBV, phorbol esters [72] and, notably, by interleukin-4 (IL-4). This latter finding is a clear example of an interaction between specialized IL-4-producing $CD4^+$ memory-type T cells [73, 74] and B lymphocytes which is likely to lead to B cell activation in the *paracortical areas* of lymph nodes [75]. This interaction is well controlled because the IL-4-mediated up-regulation of CD23 on B cells is blocked in the

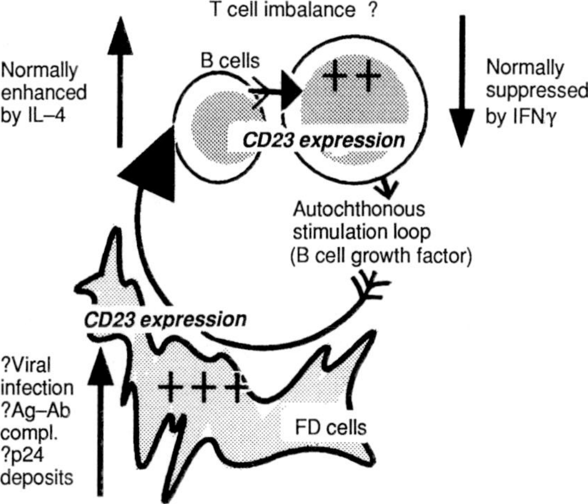

Fig. 6. Hypothesis: immunoregulatory imbalance leading to B cell hyperactivity in HIV-1⁺ lymphadenitis. IL-4 increases CD23 expression on B cells and IFNγ blocks this action. The CD23 molecule appears to be an autochthonous growth factor for B blast cells [75]. Thus an imbalance between the synthesis of IL-4-like substances and IFNγ by HIV-1 affected T cells (e.g. selective local IFNγ deficiency in the vicinity of activated B cells) would lead to B cell hyperactivity but this hypothesis is unproven. On the other hand, the hyperexpression of CD23 by FD cells in HIV-1⁺ lymphadenitis has been documented (see fig. 3). While the exact causes of this phenomenon are still to be investigated (T cell imbalance? HIV-1 infection of FD cells? Deposition of p24 complexes in the germinal centre?) the possibility of B cell hyperactivity on the background of moderate T cell deficiency is reinforced. Similar phenomena are frequently seen in other diseases of T cell deficiency such as Hodgkin disease, angioimmunoblastic lymphadenopathy, disregulated regeneration following bone marrow transplantation, etc.

presence of interferon-γ (IFNγ) [76], a lymphokine also secreted by subsets of $CD4^+$ cells. These observations, summarized in figure 6, emphasize that HIV-1 infection may interfere with the development of IL-4 and IFγ-secreting $CD4^+$ T cells to a differing degree and then lead to various secondary B cell irregularities such as up- and down-regulation depending upon the changing balance between the IL-4/IFNγ secretion. The development of IFNγ deficiency has indeed been demonstrated in progressing HIV-1 infection [77] and the low natural killer activity in these patients is also probably attributable to this phenomenon [78], but the interrelationships with the IL-4 secretory status are still to be studied.

The most relevant point, however, is that in lymph nodes the richest $CD23^+$ cells are seen inside the germinal centres among a specialized subset of FD cells which are normally confined to the apical light zone (fig. 2d). This is a moderately broad band of cells which presents antigens complexed to antibodies and complement to the developing centrocytes in the middle of germinal centres [79]. The first issue here is that the regulation of CD23 expression of FD cells is totally unknown. The following factors are good candidates for altering this expression: (a) the viral infection of FD cells; (b) germinal centre seeking IL-4 and IFNγ-secreting T cell populations – which may have $CD4^+$, $CD8^+$ and NK-like (Leu-7^+) phenotypes; (c) complexes of antigen-antibody complement, e.g. those which accumulate large quantities of p24 core antigens. None of these have been investigated despite their clear documentation inside the germinal centres during HIV-1^+ lymphadenitis (fig. 1–3). It is also unknown whether or not these FD cells express only membrane-bound CD23 or release these moieties as soluble CD23 and in which form [reviewed in 75]. The fact remains, however, that in the HIV-1^+ lymph nodes the bands of $CD23^+$ FD cells are enormously expanded and fill virtually the whole enlarged germinal centre. The intensity of staining is also greatly increased (fig. 2d). Even the remnants of FD cells in the involuted lymph nodes are strongly $CD23^+$ (fig. 2e).

The functional significance of these findings are twofold. Firstly, the CD23 antigen is an autochthonous B cell growth factor which may promote autostimulatory activity [68, 69]. It has always been puzzling in developing AIDS that T cell depression has been associated with B cell hyperactivity. As the dogma about the T-cell dependence of humoral immunity is frequently exaggerated, the lack of suppressor/inducer cells [13] or the existence of mitogenic stimuli, such as co-infection of B cells with EBV [14] or subliminal activation of T cells by gp120 [80] have been postulated, but these possibilities may not explain the discrepancy fully. At the same time the well-documented details about a dichotomy of the humoral responses in terms of T cell (in) dependence [81–84] have not been discussed. These classical studies have underlined that efficient T cell help is only required for the development and stimulation of B cells which synthesize antibodies of high affinity and switch their Ig classes from IgM to IgG/IgA during their development, while immunoglobulin synthesis with antibody activity of low affinity can proceed without much T help. In developing AIDS, in the same way as seen in other conditions of T cell deficiency (e.g. Hodgkin disease, etc.), hypergammaglobulinaemia may develop together with defects of making antibodies of high affinity. The documentation of CD23-rich FD cells may provide clues about the molecular mechanisms how this relatively T-cell independent B-cell hyperactivity is generated, but the exact steps leading to this CD23 overexpression are still unknown. Secondly, the simplified scheme shown in figure 6 has some features which may help in explaining the deterioration of germinal centres from a hyperactive to an involuted state. If the factors which provide the negative feedback (e.g. IFNγ) are made in insufficient amounts (due to a still minor T cell deficiency), the result might be a hyperactive germinal centre where the lack of appropriate somatic mutations in the B cell populations (i.e. the lack of generating antibodies of 'quality') are temporarily masked by producing more B cells which then make poor antibodies in 'quantity'. But when the defects run deeper and the livelihood of FD cells is jeopardized by a further deterioration (e.g. by the total lack of producing IL-4-like substances), the germinal centres involute.

References

1 Harper ME, Marselle LM, Gallo RC, et al: Detection of lymphocytes expressing human T-lymphotropic virus type III in lymph nodes and peripheral blood from infected individuals by in situ hybridization. Proc Natl Acad Sci USA 1986;83:772–776.

2 Biberfeld P, Chayt KJ, Marselle LM, et al: HTLV-III expression in infected lymph nodes and relevance to pathogenesis of lymphadenopathy. Am J Pathol 1986;125:436–442.

3 Tenner Racz K, Racz P, Schmidt H, et al: Immunohistochemical, electron microscopic and in situ hybridization evidence for the involvement of lymphatics in the spread of HIV-1. Aids 1988;2:299–309.

4 Schuurman HK, Krone WJA, Broekhuizen R, et al: Expression of RNA and antigens of human immunodeficiency virus type-1 (HIV-1) in lymph nodes from HIV-1 infected individuals. Am J Pathol 1988;133:62–70.

5 Wiley CA, Schrier RD, Nelson JA, et al: Cellular localization of human immunodeficiency virus infection within the brains of acquired immune deficiency syndrome patients. Proc Natl Acad Sci USA 1986;83:7089–7093.

6 Singer RH: Use of non-radioactive labelled probes; in Racz P (ed): Molecular Pathology of Retroviral Infections and AIDS. Basel, Karger, 1989.

7 Psallidopoulos M, Schnittman S, Baseler M, et al: CD4 positive T lymphocytes from PBMC of healthy seropositive individuals harbour HIV-1 proviral sequences (abstract). Vth Int Conf AIDS 1989;5:510.

8 Spear GT, Moore JL, Landay A, et al: Monocytes and neutrophils from HIV-infected individuals contain very low or undetectable levels of provirus as determined by PCR analysis (abstract). Vth Int Conf AIDS 1989;5:570.

9 Gruters RA, Noesel CJMV, Terpstra FG, et al: Selective signaling defect via the CD3/TCR complex in T cells from HIV infected men (abstract). Vth Int Conf AIDS 1989;5:533.

10 Hornicek FJ, Malinin GI, Thornthwaite JT, et al: Effet of mitogens on the cell cycle progression and the quantification of T-lymphocyte surface markers in acquired immune deficiency syndrome. J Leukocyte Biol 1987;42:122–127.

11 Duesberg P: HIV is not the cause of AIDS. Science 1988;241:514–517.

12 Katz JD, Nishanian P, Mitsuyasu R, et al: Antibody-dependent cellular cytotoxicity (ADCC)-mediated destruction of human immunodeficiency virus (HIV)-coated $CD4^+$ T lymphocytes by acquired immunodeficiency syndrome (AIDS) effector cells. J Clin Immunol 1988;8:453–458.
13 Nicholson JK, Jones BM, Echenberg DF, et al: Phenotypic distribution of T cells of patients who have subsequently developed AIDS. Clin Immunol Immunopathol 1987;43:82–87.
14 Lane HC, Fauci AS: Immunologic abnormalities in the acquired immunodeficiency syndrome. Annu Rev Immunol 1985;3:477–500.
15 Pileri S, Rivano MT, Raise E, et al: The value of lymph node biopsy in patients with the acquired immunodeficiency syndrome (AIDS) and the AIDS-related complex (ARC): a morphological and immunohistochemical study of 90 cases. Histopathology 1986;10:1107–1129.
16 Pallesen G, Gerstoft J, Mathiesen L: Stages in LAV/HTLV-III lymphadenitis. I. Histological and immunohistological classification. Scand J Immunol 1987;25:83–91.
17 Tenner Racz K, Racz P, Dierich M, et al: Monoclonal antibodies to HIV: their relation to the pattern of changes in PGL and AIDS. Aids 1987;1:95–104.
18 Seligmann M, Pinching AJ, Rosen FS, et al: Immunology of human immunodeficiency virus infection and the acquired immunodeficiency syndrome. An update. Ann Intern Med 1987;107:234–242.
19 Grody WW, Fliegel S, Haeim F: Thymic involution in the acquired immunodeficiency syndrome. Am J Clin Pathol 1985;84:85–95.
20 Delacretaz F, Perey L, Schmidt PM, et al: Histopathology of bone marrow in human immunodeficiency virus infection. Virchows Arch (A) 1987;411:543–551.
21 Rousselet MC, Audouin J, LeTourneau A, et al: Idiopathic thrombocytopenic purpura in patients at risk for acquired immunodeficiency. Histopathologic study, immunohistochemistry, and ultrastructural study on six spleens. Arch Pathol Lab Med 1988;112:1242–1250.
22 Bishop PE, McMillan A, Gilmour HM: Immunological study of the rectal mucosa of men with and without human immunodeficiency virus infection. Gut 1987;28: 1619–1624.
23 Vazeux R, Brousse N, Jarry A, et al: AIDS subacute encephalitis. Identification of HIV-infected cells. Am J Pathol 1987;126:403–410.
24 Janossy G, Bofill M, Poulter LW, et al: Separate ontogeny of two macrophage-like accessory cell populations in the human fetus. J Immunol 1986;136:4354–4360.
25 Parwaresch MR, Radzun HJ, Hansmann ML, et al: Monoclonal antibody Ki-M4 specifically recognizes human dendritic reticulum cells (follicular dendritic cells) and their possible precursor in blood. Blood 1983;62:585–590.
26 Wood GS, Burns BF, Dorfman RF, et al: The immunohistology of non-T cells in the acquired immunodeficiency syndrome. Am J Pathol 1985;120:371–379.
27 Racz P, Tenner Racz K, Kahl C, et al: Spectrum of morphologic changes of lymph nodes from patients with AIDS or AIDS-related complexes. Prog Allergy 1986;37:81–181.
28 Knapp W: Leucocyte typing IV. 1989;4:in press.
29 O'Garra A, Umland S, De France T, et al: 'B cell factors' are pleiotropic. Immunol Today 1988;9:45–54.
30 Vago L, Concetta AM, Silvia C, et al: Morphogenesis, evolution and prognostic significance of lymphatic tissue lesions in HIV infection. Appl Pathol., in press.

31 Nakamura S, Salahuddin SZ, Biberfeld P, et al: Kaposi's sarcoma cells: long-term culture with growth factor from retrovirus-infected $CD4^+$ T cells. Science 1988;242:426–430.
32 Tenner-Racz K, Racz P, Bofill M, et al: HTLV III/LAV viral antigens in lymph nodes of homosexual men with persistent generalized lymphadenopathy and AIDS. Am J Pathol 1987;123:9–15.
33 Janossy G, Pinching AJ, Bofill M, et al: An immunohistological approach to persistent lymphadenopathy and its relevance to acquired immune deficiency syndrome. Clin Exp Immunol 1985;59:257–266.
34 Wood GS, Burns BF, Dorfman RF, et al: In situ quantitation of lymph node helper, suppressor, and cytotoxic T cell subsets in AIDS. Blood 1986;67:596–603.
35 Brask S, Hager H, Pallesen G, et al: Quantification of CD8-positive lymphocytes in lymph node follicles from HIV-infected male homosexuals and controls. Acta Pathol Microbiol Immunol Scand (A) 1987;95:155–157.
36 Racz P: Molecular, biologic, immunohistochemical, and ultrastructural aspects of lymphatic spread of the human immunodeficiency virus. Lymphology 1988;21:28–35.
37 Warner NL: Human lymphocyte subpopulations: analysis by multiparameter flow cytometry and monoclonal antibodies. Cancer Detect Prev 1987(suppl 1):515–523.
38 Caruso A, Terlenghi L, Scalzini A, et al: Evaluation of the expression of IFN-gamma in lymphocytes using a monoclonal antibody and flow cytometry. J Immunol Methods 1988;113:37–43.
39 Chilosi M, Menestrina F, Capelli P, et al: Immunohistochemical analysis of sarcoid granulomas. Evaluation of $Ki67^+$ and interleukin-1^+ cells. Am J Pathol 1988;131:191–198.
40 Hynes RO: Integrins: a family of cell surface receptors. Cell 1987;48:681–690.
41 Inoue A, Yanagisawa M, Kimura S, et al: The human endothelin family: three structurally and pharmacologically distinct isopeptides predicted by three separate genes. Proc Natl Acad Sci USA 1989;86:2863–2867.
42 Morimoto C, Letvin NL, Boyd AW, et al: The isolation and characterization of the human helper inducer T cell subset. J Immunol 1985;134:3762–3769.
43 Hemler ME, Jacobson JG: Cell matrix adhesion-related proteins VLA-1 and VLA-2: regulation of expression on T cells. J Immunol 1987;138:2941–2948.
44 Tedder TF, Cooper MD, Clement LT: Human lymphocyte differentiation antigens HB-10 and HB-11. II. Differential production of B cell growth and differentiation factors by distinct helper T cell subpopulations. J Immunol 1985;134:2989–2994.
45 Morimoto C, Letvin NL, Distaso JA, et al: The isolation and characterization of the human suppressor inducer T cell subset. J Immunol 1985;134:1508.
46 Smith SH, Brown MH, Rowe D, et al: Functional subsets of human helper-inducer cells defined by a new monoclonal antibody, UCHL1. Immunology 1986;58:63.
47 Terstappen LWMM, Loken MR: Five-dimensional flow cytometry as a new approach for blood and bone marrow differentials. Cytometry 1988;9:548–556.
48 Brenner MB, McLean J, Schaft H, et al: Characterization and expression of the human α/β T cell receptor by using a framework monoclonal antibody. J Immunol 1987;138:1502.
49 Band H, Hochstenbach F, McLean J, et al: Immunochemical proof that a novel rearranging gene encodes the T cell receptor λ subunit. Science 1987;238:682.
50 Abo T, Miller CA, Balch CM: Characterization of human granular lymphocyte subpopulations expressing HNK-1 (Leu-7) and Leu-11 antigens in the blood and lymphoid tissues from fetuses, neonates and adults. Eur J Immunol 1984;14:616–623.

51 Lanier LL, Benike CJ, Phillips JH, et al: Recombinant interleukin 2 enhanced natural killer cell-mediated cytotoxicity in human lymphocyte subpopulations expressing the Leu7 and Leu11 antigens. J Immunol 1985;134:794–801.
52 Gatenby PA, Kansas GS, Chen XY, et al: Dissection of immunoregulatory subpopulations of T lymphocytes within the helper and suppressor sublineages in man. J Immunol 1982;129:1997–2000.
53 Nadler LM: B Cell/Leukemia Panel Workshop; in Reinherz EL, Haynes BF, Nadler LM, et al (eds): Leukocyte Typing II. Berlin, Springer, 1986, vol 2, pp 3–43.
54 Rappersberger K, Gartner S, Schenk P, et al: Langerhans' cells are an actual site of HIV-1 replication. Intervirology 1988;29:185–194.
55 Gerdes J, Lemke H, Baisch H, et al: Cell cycle analysis of a cell proliferation-associated human nuclear antigen defined by the monoclonal antibody Ki-67. J Immunol 1984;133:1710–1715.
56 Biberfeld P, Parravicini C, Petren AL, et al: CNS and lymph node changes in HIV infection (abstract). Vth Int Conf AIDS 1989;5:603.
57 Clayton LK, Sieh M, Pious DA, et al: Identification of human CD4 residues affecting class II MHC versus HIV-1 gp120 binding. Nature 1989;339:548–551.
58 Raff MC, Cantor H: Subpopulations of thymus cells and thymus-derived cells. Prog Immunol 1971;1:83–88.
59 Akbar AN, Bortwick N, Bofill M, et al: Sequential changes of cytotoxic and phenotypic features of CD8 positive cells in HIV-1 infected individuals, in press.
60 Akbar AN, Terry L, Timms A, et al: Loss of CD45R and gain of UCHL1 reactivity is a feature of primed T cells. J Immunol 1988;140:2171–2178.
61 Sanders ME, Makgoba MW, Sharrow SO, et al: Human memory T lymphocytes express increased levels of three cell adhesion molecules (LFA-3, CD2 and LFA-1) and three other molecules (UCHL1, CDw29 and Pgp-1) and have enhanced IFN-gamma production. J Immunol 1988;140:1401–1407.
62 Janossy G, Bofill M, Rowe D, et al: The tissue distribution of T lymphocytes expressing different CD45 polypeptides. Immunology 1988;66:517–525.
63 Janossy G, Campana D, Akbar A: Kinetics of T lymphocyte development. Curr Top Pathol 1989;79:59–99.
64 Janossy G, Bofill M, Racz P: Changes in the organization of follicular dendritic cells in HIV-1 infected lymph nodes, in press.
65 Walker BD, Chakrabarti S, Moss B, et al: HIV-specific cytotoxic T lymphocytes in seropositive individuals. Nature 1987;328:345–348.
66 Nixon DF, Townsend AR, Elvin JG, et al: HIV-1 gag-specific cytotoxic T lymphocytes defined with recombinant vaccinia virus and synthetic peptides. Nature 1988;336:484–487.
67 Moran P, Ledbetter J, Sias J, et al: HIV-infected humans, but not chimpanzees, have cytotoxic T lymphocytes that lyse uninfected $CD4^+$ cells (abstract). Vth Int Conf AIDS 1989;5:601.
68 Kintner C, Sugden B: Identification of antigenic determinants unique to the surfaces of cells transformed by Epstein-Barr virus. Nature 1981;294:458–460.
69 Gordon J, Rowe M, Walker L, et al: Ligation of the CD23, p45 (BLAST-2, EBVCS) antigen triggers the cell-cycle progression of activated B lymphocytes. Eur J Immunol 1986;16:1075–1080.
70 Yukawa K, Kikutani H, Owaki H, et al: A B cell-specific differentiation antigen, CD23, is a receptor for IgE (Fc epsilon R) on lymphocytes. J Immunol 1987;138:2576–2580.

71 Swendeman SL, Thorley Lawson D: Soluble CD23/BLAST-2 (S-CD23/Blast-2) and its role in B cell proliferation. Curr Top Microbiol Immunol 1988;141:157–164.
72 Calender A, Billaud M, Aubry JP, et al: Epstein-Barr virus (EBV) induces expression of B-cell activation markers on in vitro infection of EBV-negative B-lymphoma cells. Proc Natl Acad Sci USA 1987;84:8060–8064.
73 Bonnefoy JY, Defrance T, Peronne C, et al: Human recombinant interleukin 4 induces normal B cells to produce soluble CD23/IgE-binding factor analogous to that spontaneously released by lymphoblastoid B cell lines. Eur J Immunol 1988;18:117–122.
74 Lewis DB, Prickett KS, Larsen A, et al: Restricted production of interleukin 4 by activated human T cells. Proc Natl Acad Sci USA 1988;85:9743–9747.
75 Gordon J, Flores-Romo L, Cairns JA, et al: CD23: a multi-functional receptor/lymphokine?. Immunol Today 1989;10:153–157.
76 Paliard X, de Waal Malefijt R, Yssel H, et al: Simultaneous production of IL-2, IL-4, and IFN-gamma by activated human $CD4^+$ and $CD8^+$ T cell clones. J Immunol 1988;141:849–855.
77 Murray HW, Scavuzzo DA, Kelly CD, et al: $T4^+$ cell production of interferon-gamma and the clinical spectrum of patients at risk for and with acquired immunodeficiency syndrome. Arch Intern Med 1988;148:1613–1616.
78 Kedar E, Rezai AR, Giorgi JV, et al: Immunomodulating effects in vitro of interleukin-2 and interferon-gamma on human blood and bone marrow mononuclear cells. Nat Immun Cell Growth Regul 1988;7:13–30.
79 Johnson GD, HArdie DL, Ling NR, et al: Human follicular dendritic cells (FDC): a study with monoclonal antibodies (MoAb). Clin Exp Immunol 1986;64:205–213.
80 Ascher M, Sheppard H: AIDS – panergic amnesia as pathogenic mechanism (abstract). Vth Int Conf AIDS 1989;5:676.
81 Humphrey JH, White RH: Immunology for Students. 3rd Ed. Oxford, Blackwell Scientific, 1974.
82 Klaus GGB, Kunkl A: The role of germinal centres in the generation of immunological memory. Ciba Found Symp 1984;84:256–280.
83 Amlot PL, Hayes AE: Impaired human antibody response to the thymus independent antigen DNP-Ficoll after splenectomy. Lancet 1985;i:1008–1011.
84 Lane HC, Depper JM, Greene WC, et al: Qualitative analysis of immune function in patients with the acquired immunodeficiency syndrome. Evidence for a selective defect in soluble antigen recognition. N Engl J Med 1985;313:79–84.

Prof. George Janossy, MD, Department of Immunology, Royal Free Hospital School of Medicine, Rowland Hill Street, London NW3 2PF (UK)

Subject Index